Lecture Notes in Computer Science　　12643

More information about this subseries at http://www.springer.com/series/7412

Anja Hennemuth · Leonid Goubergrits ·
Matthias Ivantsits · Jan-Martin Kuhnigk (Eds.)

Cerebral Aneurysm Detection and Analysis

First Challenge, CADA 2020
Held in Conjunction with MICCAI 2020
Lima, Peru, October 8, 2020
Proceedings

Springer

Editors
Anja Hennemuth 🆔
Charité – Universitätsmedizin Berlin
Berlin, Germany

Leonid Goubergrits 🆔
Charité – Universitätsmedizin Berlin
Berlin, Germany

Matthias Ivantsits
Charité – Universitätsmedizin Berlin
Berlin, Germany

Jan-Martin Kuhnigk
Fraunhofer Institute for Digital
Medicine MEVIS
Berlin, Germany

ISSN 0302-9743   ISSN 1611-3349   (electronic)
Lecture Notes in Computer Science
ISBN 978-3-030-72861-8   ISBN 978-3-030-72862-5   (eBook)
https://doi.org/10.1007/978-3-030-72862-5

LNCS Sublibrary: SL6 – Image Processing, Computer Vision, Pattern Recognition, and Graphics

This Springer imprint is published by the registered company Springer Nature Switzerland AG
The registered company address is: Gewerbestrasse 11, 6330 Cham, Switzerland

Preface

Welcome to the proceedings of the Cerebral Aneurysm Detection and Analysis (CADA) challenge, which was held as a satellite event at the 23rd International Conference on Medical Image Computing and Computer Assisted Intervention (MICCAI 2020). The CADA challenge was scheduled to take place in Lima, Peru, on October 8, 2020, but was instead held through a virtual conference management platform due to the COVID-19 pandemic.

Cerebral aneurysms are local dilations of arterial blood vessels caused by a weakness of the vessel wall. Subarachnoid hemorrhage (SAH) caused by a cerebral aneurysm rupture is a life-threatening condition associated with high mortality and morbidity. It is therefore highly desirable to detect aneurysms early and determine the appropriate rupture prevention strategy. Diagnosis and treatment planning are based on angiographic imaging using MRI, CT, or X-ray rotation angiography. Primary goals in image analysis are the detection, segmentation, and risk assessment of aneurysms using geometric parameters alone or combined with hemodynamic parameters, usually assessed by image-based computational fluid dynamics (CFD) analysis.

We, therefore, proposed a challenge with three categories with increasing challenge complexity level. The first task was detecting the aneurysm; the second task was accurate segmentation to allow for a longitudinal assessment of suspicious aneurysms' development; and the third, and most challenging, task was the estimation of the aneurysm rupture risk.

The CADA imaging database was acquired through digital subtraction angiography (DSA) utilizing the AXIOM Artis C-arm system with a rotational acquisition time of 5 s with 126 frames. A contrast agent (Imeron 300, Bracco Imaging Deutschland GmbH, Germany) was manually injected into the internal carotid (anterior aneurysms) or vertebral (posterior aneurysms) artery. In total, the database consists of 131 rotational X-ray angiographic image datasets from different patients. Each sample contains an accurate delineation of the cerebral aneurysm annotated by an expert with a minimum of five years experience, the patients' age and sex, and the rupture status. A private dataset of 22 images was held back as a test set by the CADA challenge organizers during the training period and only later released to allow offline processing and the submission of results for centralized evaluation. The remaining 109 cases were available for model training and validation.

The CADA challenge was designed to present an open and fair platform for various research groups to test and validate their methods on datasets acquired from the clinical environment. The main aims were to benchmark various cerebral aneurysm detection, segmentation, and rupture risk estimation algorithms and to generate focused discussions, especially between groups working on CFD and data-driven approaches that, perhaps, could be combined to provide more comprehensive solutions.

The results submitted for the detection and segmentation sub-challenges presented solutions which perform similarly to human experts. The top-performing solutions submitted for the rupture risk estimation sub-challenge utilized a combination of morphological and CFD features. Additionally, one of the methods extended these features with features learned by a convolutional neural network.

The nine full papers presented in this volume were selected from a multitude of methods for which results had been submitted. In addition to the papers describing solutions for the challenge tasks, we accepted two contributions that address the clinical background and the state of the art along with the methodology of the challenge. All paper submissions were carefully reviewed and revised. More information about the CADA challenge can also be found at the following website: https://cada.grand-challenge.org/.

We would like to sincerely thank the CADA challenge committee members, the participants, the MICCAI 2020 organizers, the reviewers, and our sponsors for their time, efforts, contributions, and support.

March 2021   Anja Hennemuth
                                                                                          Leonid Goubergrits
                                                                                          Matthias Ivantsits
                                                                                          Jan-Martin Kuhnigk

---

The original version of the book was revised: The book title has been changed. The correction to the book is available at https://doi.org/10.1007/978-3-030-72862-5_12

Organization

Program Chairs

Anja Hennemuth   Charité - Universitätsmedizin Berlin
Leonid Goubergrits   Charité - Universitätsmedizin Berlin
Matthias Ivantsits   Charité - Universitätsmedizin Berlin

Contents

Overview of the CADA Challenge at MICCAI 2020

Cerebral Aneurysm Detection and Analysis Challenge 2020 (CADA)

Matthias Ivantsits[1]([✉]), Leonid Goubergrits[1,6]([✉]), Jan-Martin Kuhnigk[2],
Markus Huellebrand[1,2], Jan Brüning[1], Tabea Kossen[1], Boris Pfahringer[1],
Jens Schaller[1], Andreas Spuler[5], Titus Kuehne[1,3,4], and Anja Hennemuth[1,2,3,4]

[1] Charité – Universitätsmedizin Berlin, Augustenburger Pl. 1,
13353 Berlin, Germany
{matthias.ivantsits,leonid.goubergrits}@charite.de
[2] Fraunhofer MEVIS, Am Fallturm 1, 28359 Bremen, Germany
[3] German Heart Institute Berlin, Augustenburger Pl. 1, 13353 Berlin, Germany
[4] DZHK (German Centre for Cardiovascular Research), Berlin, Germany
[5] Helios Hospital Berlin-Buch, Schwanebecker Chaussee 50, 13125 Berlin, Germany
[6] Einstein Center Digital Future, Wilhelmstraße 67, 10117 Berlin, Germany

Abstract. Rupture of an intracranial aneurysm often results in subarachnoid hemorrhage, a life-threatening condition with high mortality and morbidity. The Cerebral Aneurysm Detection and Analysis (CADA) competition was organized to support the development and benchmarking of algorithms for the detection, analysis, and risk assessment of cerebral aneurysms in X-ray rotational angiography (3DRA) images. 109 anonymized 3DRA datasets were provided for training, and 22 additional datasets were used to test the algorithmic solutions. Cerebral aneurysm detection was assessed using the F2 score based on recall and precision, and the fit of the delivered bounding box was assessed using the distance to the aneurysm. Segmentation quality was measured using Jaccard and a combination of different surface distance measurements. Systematic errors were analyzed using volume correlation and bias. Rupture risk assessment was evaluated using the F2 score. 158 participants from 22 countries registered for the CADAchallenge. The detection solutions presented by the community are mostly accurate (F2 score 0.92) with a small number of missed aneurysms with diameters of 3.5 mm. In addition, the delineation of these structures is very good with a Jaccard score of 0.915. The rupture risk estimation methods achieved an F2 score of 0.7. The performance of the detection and segmentation solutions is equivalent to that of human experts. In rupture risk estimation, the best results are obtained by combining different image-based, morphological and computational fluid dynamic parameters using machine learning methods.

Keywords: Intracranial aneurysms · Subarachnoid hemorrhage · X-ray rotational angiography · Machine learning · Rupture risks · Deep learning

© Springer Nature Switzerland AG 2021
A. Hennemuth et al. (Eds.): CADA 2020, LNCS 12643, pp. 3–17, 2021.
https://doi.org/10.1007/978-3-030-72862-5_1

1   Introduction

Cerebral aneurysms, also known as intracranial aneurysms, are local dilatations of blood vessels caused by vessel wall weakness. Aneurysms occur at different positions in the Circle of Willis and are usually classified by shape, size, and phenotype [8]. The size is determined via the diameter, and grouped into small (less than 5 mm), and middle sized-sized (5 to 15 mm), large (15 to 25 mm), giant (25 to 50 mm), and super-giant (over 50 mm) aneurysms [9]. Shape-wise, these aneurysms are categorized by the literature into saccular, fusiform, and microaneurysms. Saccular aneurysms—also known as berry aneurysms—appear as round dilatations of the vessel and are the most common form [10]. Fusiform aneurysms represent a widening of a whole segment of the vessel and are usually not at risk of rupturing but cause severe health issues. Microaneurysms—also known as Charcot–Bouchard aneurysms—usually emerge in small vessels with less than 300 μm in diameter and are a common cause of intracranial hemorrhage. Figure 1 illustrates two exemplary images with highlighted aneurysms. The primary risk associated with aneurysms is rupture, causing subarachnoid hemorrhage (SAH). SAH is a life-threatening condition with high mortality and morbidity [6,7]. The death rate is above 40% [11], and in case of survival, cognitive impairment can affect patients for a long time, even lifelong. Therefore, it is highly desirable to detect unruptured aneurysms early and decide about the appropriate rupture prevention strategy, including the decision between treatment and observation and the treatment procedure's choice. Visual detection of small aneurysms is challenging considering the complex vessel structure of the Circus Willis, half of which is well shown in Fig. 1, right. Taking into account the above-described challenges for clinicians as well as a history of other aneurysm challenges described below in the next chapter, we elaborated the CADA challenge with three subtasks:

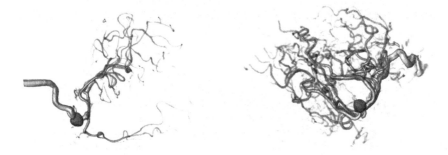

Fig. 1. Volume rendering visualization of two exemplary rotational X-ray angiographic images with highlighted cerebral aneurysms in red. (Color figure online)

1.1   Task 1: Aneurysm Detection

Automatic detection of unknown aneurysms in the brain imaging of patients with or without symptoms and detection of unruptured aneurysms in patients with known ruptured aneurysms can improve patient outcome and save patients' lives.

1.2   Task 2: Aneurysm Segmentation

Segmentation of aneurysms together with parent vessels is the first step for the treatment decision support allowing the assessment of the shape and size of the aneurysms associated with the rupture risk. Furthermore, segmentation followed by geometry reconstruction is the first step for the image-based simulations of the individual hemodynamics associated with the rupture risk.

1.3   Task 3: Aneurysm Rupture Risk Estimation

Automatic classification of known ruptured and unruptured aneurysms is the first step towards the development of the decision support system identifying unruptured aneurysms prone to rupture.

The CADA challenge was organized at the 23rd International Conference on Medical Image Computing and Computer-Assisted Intervention 2020 (MICCAI). The brief challenge description can be found here: [12]. The challenge's primary objective was to evaluate variability and respectively current state of the art of deep learning solutions for the three above described tasks. We provide an overview of the results achieved with the different algorithms submitted to the challenge.

2   Dataset

The CADA dataset [12] was acquired utilizing the digital subtraction AXIOM Artis C-arm system with a rotational acquisition time of 5 s with 126 frames (190° or 1.5° per frame, 1024 × 1024-pixel matrix, 126 frames). Post-processing was performed using LEONARDO InSpace 3D (Siemens, Forchheim, Germany). A contrast agent (Imeron 300, Bracco Imaging Deutschland GmbH, Germany) was manually injected into the internal carotid (anterior aneurysms) or vertebral (posterior aneurysms) artery. Reconstruction of a volume of interest selected by a neurosurgeon generated a stack of 440 image slices with matrices of 512 × 512 voxels in-plane, resulting in an iso-voxel size of 0.25 mm. The images were acquired in the Neurosurgery Department, Helios Klinikum Berlin-Buch. In total, the dataset consists of 131 rotational X-ray angiographic images from different patients. A private dataset of 22 images was held back as a test set by the CADA challenge organizers during the training period and only later released to allow offline processing and send in the results for centralized evaluation. The remaining 109 cases were available for model training and validation.

Figure 2 illustrates the demographics distribution of the patients, with an imbalance in the sex distribution in the training and the test set, but showing similar distributions across the two sets. Females in the training set are, on aver-

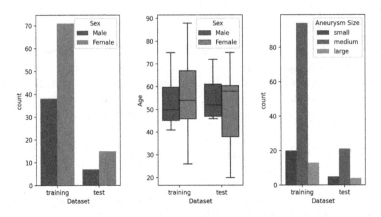

Fig. 2. The patients sex distribution on the left. The patients age distribution in the center for the training and test set. On the right the aneurysm size distribution by dataset.

age, 54 ± 13.6 years, with males being 54 ± 10.5 years old. A similar distribution for the test set with 51 ± 16 years for women and 55 ± 10 for men.

Figure 3 illustrates the number of aneurysms per patient and the diameter of the vessel dilations for the training and test set, respectively. Most patients in both sets have one aneurysm; roughly 10% of the datasets show more than one cerebral aneurysm. The center plot in Fig. 3 highlights the diameters of the aneurysms in the two sets by sex. The training set reveals a mean structure size of 9 ± 5 for females and 8.7 ± 3.8 for males. The test set contains aneurysms in diameter of 9 ± 6.3 for women and 10.1 ± 6.1 for men. The plot on the right side of Fig. 3 illustrates the sphericity of the aneurysms, which describes the overall roundness of a structure relative to a sphere. The value range is $0 < sphericity \leq 1$, where a value of 1 indicates a perfect sphere (the formula is given by Eq. 1). A stands for the surface area of a given structure and V for the volume. This analysis implies that the datasets contain mainly small-sized saccular aneurysms.

$$sphericity = \frac{\sqrt[3]{36\pi V^2}}{A} \qquad (1)$$

Equation 1: The sphericity of a given structure, where A stands for the surface area of a given structure and V for the volume.

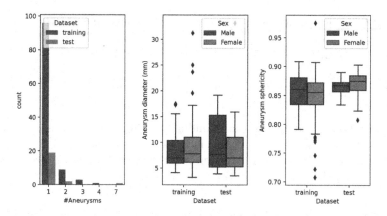

Fig. 3. The number of aneurysms per patient for the two datasets on the left. In the center the aneurysm diameter distribution and the sphericity distribution on the right.

3   Methods

3.1   Evaluation Metrics

Detection. The major goal in detection is to ensure that aneurysms, which may pose a stroke risk, are not overlooked, so sensitivity is an important measure. On the other hand, if the whole image is marked, the aneurysms would be included, but the information would be meaningless, so precision is important as well. A bounding box is helpful if it supports the visualization and post-processing. To this end, it should contain the aneurysm but be as small as possible. The ranking is based on the F2-score (Eq. 2) that combines recall and precision, considering recall twice as important as precision. The bounding box metrics (Eqs. 3, 4) is only used in case of an equal ranking. For the ranking, we perform a normalization according to the maximum among all participants. Each metric takes a value between 0 (worst case among all participants) and 1 (perfect fit between the reference and predicted segmentation). The ranking score is calculated as the average of the normalized metrics.

$$F_2 = 5\frac{P \cdot R}{(4 \cdot P) + R} \tag{2}$$

Equation 2: P is denoted as precision $P = \frac{TP}{TP+FP}$ and R is denoted as the recall $R = \frac{TP}{TP+FN}$. TP are the true-positives, FP the false-negatives, and FN the false-negatives.

$$C_{cA} = \frac{1}{|A|} \sum_{c\,in\,C} \sum_{c_A\,in\,A} \frac{|M_{c_A} \cap BB_{c_A}|}{|M_{c_A}|} \tag{3}$$

Equation 3: C_{c_A} is denoted as the coverage of the aneurysm c_A by the bounding-box. M_{c_A} is denoted as the ground-truth segmentation of the aneurysm and BB_{c_A} as the predicted bounding-box.

$$F_{c_A} = \sum_{c \, in \, C} \sum_{c_A \, in \, A} \max_{x_{c_i} \in M_{c_A}} \left(\min_{m_{c_j} \in M_{c_A}} (|x_{c_i} + v - m_j|, |x_{c_i} + v - m_j|) \right) \quad (4)$$

Equation 4: F_{c_A} is denoted as the bounding-box fit, C as the cases, A as the aneurysms within one case c, and M_{c_A} is mask for aneurysm c_A. The bounding-box fit measures the maximum distance between the bounding box and the aneurysm along the main axes of the bounding-box.

Segmentation. For assessing the segmentation quality, we compare the submission segmentation result masks $M_{c_A}^*$ with ground truth masks M_{c_A} from the expert annotations. The segmentation is the basis for the quantitative assessment of the aneurysms. It should enable the extraction of shape and volume parameters to assess change over time or compare with decision thresholds. Therefore, the overlap and distance from reference segmentations are essential. For the assessment of volumes ($V_{c_A}^*$ noted as the submitted volume and V_{c_A} noted as the expert volume), we also analyze how well the results correlate over the cohort and if there is a bias. For the ranking, we perform a normalization according to the maximum among all participants. Each metric takes a value between 0 (worst case among all participants) and 1 (perfect fit between the reference and predicted segmentation). Furthermore, the Hausdorff distance, average distance, bias, and standard deviation are inverted since lower metric scores are considered better fits. The ranking score is calculated as the average of the normalized metrics.

$$J(M_{c_A}^*, M_{c_A}) = \frac{M_{c_A}^* \cap M_{c_A}}{M_{c_A}^* \cup M_{c_A}} \quad (5)$$

Equation 5: The Jaccard of the ground-truth M_{c_A} and predicted segmentation $M_{c_A}^*$. This metric measures the voxel overlap of M_{c_A} and $M_{c_A}^*$.

Rupture Risk Estimation. For the assessment of the rupture risk prediction, we calculate recall and precision. Same as in Task 1, an aneurysm at risk should not be misclassified, but the number of false alarms should also be low. Thus, sensitivity and precision are assessed via the F2-score (Eq. 2). The final ranking is based on the F2-score (Eq. 2).

$$HD(M_{c_A}^*, M_{c_A}) = \max \left(\max_{m_j \in M_{c_A}^*} \left(\min_{m_i \in M_{c_A}^*} (|m_j - m_i|) \right) \right) \quad (6)$$

Equation 6: The Hausdorff distance (HD) from the ground-truth M_{c_A} to predicted surface $M_{c_A}^*$. HD measures the maximum minimum distance between two surfaces.

$$AVD(M_{c_A}^*, M_{c_A}) = \frac{1}{2|M_{c_A}^*|} \sum_{m_j \in M_{c_A}^*} \min_{m_i \in M_{c_A}} |m_j - m_i|$$
$$+ \frac{1}{2|M_{c_A}|} \sum_{m_j \in M_{c_A}} \min_{m_i \in M_{c_A}^*} |m_j - m_i| \tag{7}$$

Equation 7: The average distance from the ground-truth M_{c_A} to predicted surface $M_{c_A}^*$. AVD measures the average minimum distance between two surfaces.

$$\rho(V_{c_A}^*, V_{c_A}) = \frac{cov(V_{c_A}^*, V_{c_A})}{\sigma_{V_{c_A}^*} \sigma_{V_{c_A}}} \tag{8}$$

Equation 8: The correlation between the ground-truth volume V_{c_A} and the predicted volume $V_{c_A}^*$.

$$b = \frac{1}{|\{c_A\}|} \sum_{\{c_A\}} |V_{c_A}^* - V_{c_A}| \tag{9}$$

Equation 9: The bias of the ground-truth volumes V_{c_A} and the predicted volumes $V_{c_A}^*$.

4 Submissions

At the time of the CADA workshop, we record 158 registrations from 22 different countries. A map of the registered participants is illustrated in Fig. 4. Figure 5 illustrates the submissions to the CADA challenge. In total, there were 157 submissions from China, Germany, Hong Kong, Canada, and France.

5 Results

5.1 Detection

In total, eight teams submitted solutions to the CADA detection challenge. Three of these submissions presented a technical paper describing their methodology [13–15]. An overview of this detection challenge's final ranking is illustrated in Table 1. The methodology of the final ranking is illustrated in Sect. 3.1.

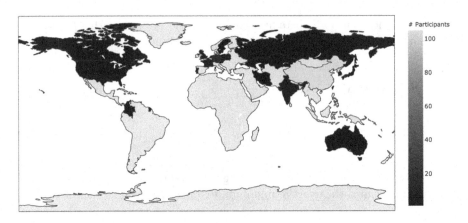

Fig. 4. An illustration of the registered participants.

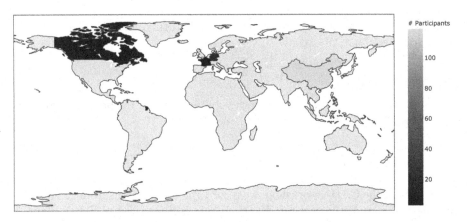

Fig. 5. An illustration of the submissions to the CADA challenge.

5.2   Segmentation

For the CADA segmentation challenge, twelve teams submitted their solutions. Three of these submissions presented a technical paper describing their methodology [14,16,17]. An overview of this segmentation challenge's final ranking is illustrated in Table 2. The methodology of the final ranking is illustrated in Sect. 3.1.

Table 1. The final ranking of the CADA detection challenge. The table includes the F2-score, coverage of the aneurysm by the bounding box, and the fit of the bounding box.

	User	F2-score	Coverage	BBoxFit
1	**Mediclouds [15]**	**0.918**	0.979	0.939
2	**ibbm [16]**	0.856	**0.989**	0.782
3	yyq, hustyyq	0.856		
4	junma	0.850	**0.989**	0.786
5	alirezasjd	0.839		
6	**jmkuhnigk [17]**	0.839		
7	chhluo	0.833	0.987	**0.705**
8	mibaumgartner	0.777	0.239	3.276

Table 2. The final ranking of the CADA segmentation challenge. The table includes the final score, Jaccard, Hausdorff distance (HD), average distance (AVD), the volume correlation, the absolute volume bias, the volume bias, and the volume standard deviation.

	User	Score	Jaccard	Distance[mm]		Corr	Volume[mm^3]		
				HD	AVD		Abs.Bias	Bias	Std
1	**Mediclouds [18]**	0.833	0.76	**2.87**	**1.62**	**0.998**	**72.24**	-2.04	**106.41**
2	**junma [19]**	0.832	**0.76**	4.97	3.54	**0.998**	75.84	4.04	110.53
3	chhluo	0.819	0.71	4.94	3.48	**0.998**	79.73	23.60	109.81
4	hezf	0.796	0.66	31.70	6.13	0.996	103.35	-28.13	129.71
5	**ibbm [16]**	0.793	0.68	38.76	36.83	0.997	99.08	-34.85	135.29
6	yyq, hustyyq	0.786	0.72	66.40	40.35	0.992	171.23	-104.57	201.51
7	alirezasjd	0.775	0.58	4.44	2.50	**0.998**	108.26	50.91	211.08
8	zhangyong7630	0.774	0.56	4.72	2.49	0.997	109.54	55.81	154.00
9	joagh96	0.768	0.58	41.78	39.70	0.997	87.24	-17.94	127.71
10	jmkuhnigk	0.705	0.63	36.72	35.05	0.879	319.26	-260.21	1334.90
11	Sssplendid	0.601	0.38	247.68	241.44	0.988	272.47	-37.68	438.11
12	YiyiYang	0.578	0.43	401.58	400.48	0.980	162.41	92.61	349.24

5.3   Rupture Risk Estimation

For the CADA rupture risk estimation challenge, three teams submitted their solutions. Two of these submissions presented a technical paper describing their methodology [18,19]. An overview of the final ranking for this rupture risk estimation challenge is illustrated in Table 3. The methodology of the final ranking is illustrated in Sect. 3.1.

Table 3. The final ranking of the CADA rupture risk estimation challenge.

	User	F2-score
1	**matthiasivantsits2 [20]**	**0.702**
2	**yyq, hustyyq [21]**	0.678
3	MarioViti2	0.377

6   Discussion and Conclusion

Aneurysm Detection worked well for large- and medium-sized aneurysms with a diameter larger than 5 mm. The best detection solution showed perfect sensitivity on these medium- and large-sized structures. For small-sized cerebral aneurysms, the best solution achieves a sensitivity of 0.9. A qualitative examination of the missed aneurysms shows that all evaluated models missed the same aneurysms. Figure 6 illustrates these missed vessel dilations (red bounding box). Both aneurysms have a diameter of ~3.5 mm and a volume of ~20 mm³. The aneurysm in Fig. 6 B is very close to another aneurysm, which makes it hard to detect as a distinct dilation. Furthermore, the aneurysm in Fig. 6 A is located in the center of an arc and therefore hard to detect for the described models. Because the detection of small aneurysms is essential in clinical routine, it is important to focus future efforts on a solution to this specific problem. Other publications verify these conclusions [20–24], showing a similar sensitivity and difficulty in the detection of small structures.

Aneurysm Segmentation. Considering the Jaccard metric, the segmentation works well but not perfect. The Jaccard inspection by aneurysm size reveals that each participant's mean results are skewed towards smaller structures. The reason for that is that the presented Jaccard measures are the arithmetic mean rather than the mean weighted by the aneurysms' volume. This becomes even clearer after the qualitative check of the presented segmentations in Fig. 7. For the top presented solutions, the majority of the error in delineating cerebral aneurysms happens around the surface of the dilation, especially towards the vessel tree (the region usually called the aneurysm neck), where the definition of the separation between vessel wall and aneurysm is difficult. This results in

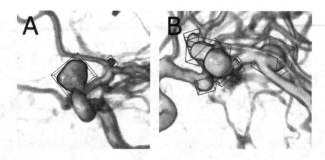

Fig. 6. An exemplary illustration of two small-sized aneurysms missed (red bounding box) by all participants. (Color figure online)

low Jaccard metrics for tiny aneurysms due to a lower surface-to-volume ratio. After the correction to this weighted mean, the results for the three published solutions look much more promising. Ma et al. [17] (junma) achieved a Jaccard of 0.915, Su et al. [16] (Mediclouds) 0.911, and Shit et al. [14] (ibbm) 0.885. These results correspond to published state-of-the-art solutions [25, 26] dealing with the segmentation of intracranial aneurysms. After a closer inspection of the volume bias in Table 2, it seems that the solution on rank six underestimates the volume systematically. An elaborate post-processing strategy might improve the results of these segmentation methods. Similar is the analysis of the method on rank ten. At first glance, it seems like an identical systematic error. Indeed, after a closer examination of the bias and standard deviation and the individual predictions, it becomes clear that one large aneurysm is the reason for this outlier. Only 20% of this aneurysm was segmented, which results in this apparent underestimation and large standard deviation. Even though the segmentation results are promising according to our assessment metrics, the impact of the deviations from the ground-truth segmentations on subsequent analysis tasks such as progression monitoring or rupture risk estimation needs to be quantified.

The Aneurysm Rupture Risk Estimation. Two of the submitted methods solved the challenge by utilizing image information only. These models are mostly based on morphological and textural radiomics features extracted from the delineated vessel dilation. A non-official solution proposed the combination of hemodynamics parameters determined via computational fluid dynamics paired with a 3D surface encoding deep neural network. Related literature [5, 27] emphasizes the high risk of rupture for small aneurysms. Interestingly, all of the proposed methods exhibit no false-negatives on small-sized aneurysms. One of the solutions presents perfect sensitivity, precision, and accuracy on these tiny structures. Ivantsits et al. [18] and Liu et al. [19] presented a solution with an F2-score of 0.7 and 0.68, respectively. Other published solutions [28, 32, 35, 36, 38–42] illustrate slightly better accuracy than the proposed methods for medium- or larger-sized aneurysms, with an accuracy maxing at around 0.8. Additional clinical parameters like smoking or hypertension [43, 44] might improve the accuracy

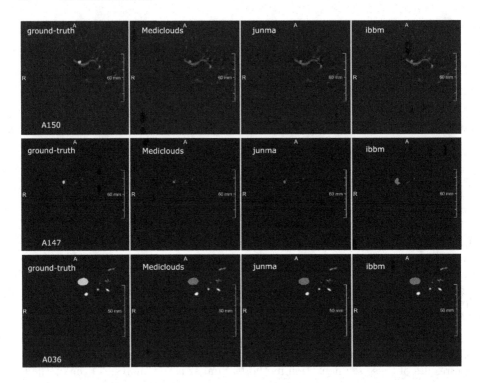

Fig. 7. A qualitative assessment of the presented segmentation models. The first column shows the ground-truth produced by an expert and the following columns highlight the segmentations inferred by the methods of the challenge participants. The first row shows a small-sized aneurysm segmentation. Su et al. [16] (Mediclouds), [14] (ibbm), and Ma et al. [17] (junma) achieve a Jaccard of 0.81, 0.72, and 0.76, as well as a Hausdorff distance of 1.1 mm, 0.7 mm, and 0.7 mm, respectively. Row two illustrates a medium-sized aneurysm, for which the provided solutions differ strongly. Su et al. [16] (Mediclouds), [14] (ibbm), and Ma et al. [17] (junma) exhibit a Jaccard of 0.47, 0.31, and 0.53, as well as a Hausdorff distance of 1.6 mm, 5.9 mm, and 1.8 mm, respectively. The last row shows a large-sized aneurysm. Su et al. [16] (Mediclouds), [14] (ibbm), and Ma et al. [17] (junma) achieve a Jaccard of 0.87, 0.82, and 0.9, as well as a Hausdorff distance of 2.4 mm, 3.0 mm, and 1.0 mm, respectively.

in predicting the rupture probability of a cerebral aneurysm. Furthermore, the implications of non-perfect segmentation of these aneurysms need to be quantified for the assessment of the uncertainty of a risk assessment.

Complementing previous challenges [1–4], the CADA challenge provides a set of datasets that also enables the evaluation of machine learning approaches. The submitted results on the recognition and segmentation sub-challenges present solutions whose performance is similar to that of human experts. State-of-the-art methods [28–37, 44] typically utilize a combination of morphological and CFD features. The top performing solutions submitted to the CADA challenge utilize similar features in combination with machine learning approaches to predict the risk of rupture. Additionally, one of the methods extends these features with features learned by a convolutional neural network.

Acknowledgement. We want to thank NVIDIA for its generous support in hosting this challenge. First of all, for their platform to execute and evaluate the participants' methods, furthermore, for GPUs' sponsorship for the top-performing solution in each sub-challenge. Moreover, we would like to thank the B. Braun-Stiftung for their benevolent sponsorship.

This work was supported by the Deutsche Forschungsgemeinschaft (DFG) under grant numbers DFG HA 5399/5-1, HE 7312/4-1, HE 1875/29-1 and the German Ministry for Education and Research (BMBF) under grant number BIFOLD-BZML (FKZ: 01IS18037E).

References

1. Valen-Sendstad, K., et al.: Real-world variability in the prediction of intracranial aneurysm wall shear stress: the 2015 international aneurysm CFD challenge. Cardiovasc. Eng. Technol. **9**, 544–564 (2018)
2. Janiga, G., et al.: The computational fluid dynamics rupture challenge 2013-phase I: prediction of rupture status in intracranial aneurysms. Am. J. Neuroradiol. **36**, 530–536 (2015)
3. Steinman, D.A., et al.: Variability of computational fluid dynamics solutions for pressure and flow in a giant aneurysm: the ASME 2012 summer bioengineering conference CFD challenge. J. Biomech. Eng. **135**, 021016 (2013)
4. Radaelli, A.G., et al.: Reproducibility of haemodynamical simulations in a subject-specific stented aneurysm model–a report on the virtual intracranial stenting challenge 2007. J. Biomech. **41**, 2069–2081 (2008)
5. Mokin, M., et al.: What size cerebral aneurysms rupture? A systematic review and meta-analysis of literature. Neurosurgery **66**, nyz310_664 (2019)
6. Morita, A., et al.: The natural course of unruptured cerebral aneurysms in a Japanese cohort. N. Engl. J. Med. **366**, 2474–2482 (2012)
7. Wiebers, D.O., et al.: Un-ruptured intracranial aneurysms: natural history, clinical outcome, and risks of surgical and endovascular treatment. Lancet **362**, 103–110 (2003)
8. Jeong, Y.-G., et al.: Size and location of ruptured in-tracranial aneurysms. J. Korean Neurosurg. Soc. **45**, 11 (2009)
9. Lasheras, J.C.: The biomechanics of arterial aneurysms. Ann. Rev. Fluid Mech. **39**, 293–319 (2007)
10. Bhidayasiri, R., et al.: Neurological Differential Diagnosis: A Prioritized Approach (2005)
11. Teunissen, L.L., et al.: Risk factors for subarachnoid hemorrhage (1996)
12. CADA Rupture Risk Estimation Challenge. https://cada-rre.grand-challenge.org/. Accessed 05 Oct 2020
13. Jia, Y., et al.: Detect and identify aneurysms based on adjusted 3D attention UNet (2021)
14. Shit, S., Ezhov, I., Paetzold, J.C., Menze, B.: Aν-net: automatic detection and segmentation of aneurysm (2021)
15. Ivantsits, M., Kuhnigk, J., Huellebrand, M., Kuehne, T., Hennemuth, A.: Deep learning-based 3D U-Net cerebral aneurysm detection (2021)

16. Su, Z., et al.: 3D attention U-Net: a solution to CADA-aneurysm segmentation challenge (2021)
17. Ma, J., Nie, Z.: Exploring large context for cerebral aneurysm segmentation (2021)
18. Ivantsits, M., Hüllebrand, M., Kelle, S., Kühne, T., Hennemuth, A.: Intracranial aneurysm rupture risk estimation utilizing vessel-graphs and machine learning (2021)
19. Liu, Y., et al.: Cerebral aneurysm rupture risk estimation using XGBoost and fully connected neural network (2021)
20. Sulayman, N., et al.: Semi-automatic detection and segmentation algorithm of saccular aneurysms in 2D cerebral DSA images. Egypt. J. Radiol. Nucl. Med. **47**, 859–865 (2016)
21. Rahmany, I., et al.: A fully automatic based deep learning approach for aneurysm detection in DSA images (2018)
22. Duan, H., et al.: Automatic detection on intracranial aneurysm from digital subtraction angiography with cascade convolutional neural networks. Biomed. Eng. Online **18**, 1–18 (2019)
23. Jin, H., et al.: Fully automated intracranial aneurysm detection and segmentation from digital subtraction angiography series using an end-to-end spatiotemporal deep neural network. J. NeuroInterventional Surg. **12**, 1023–1027 (2020)
24. Zeng, Y., et al.: Automatic diagnosis based on spatial information fusion feature for intracranial aneurysm. IEEE Trans. Med. Imaging **39**, 1448–1458 (2020)
25. Dakua, S.P., Abinahed, J., Al-Ansari, A., et al.: A PCA-based approach for brain aneurysm segmentation. Multidimens. Syst. Signal Process. **29**, 257–277 (2018)
26. Patel, T., et al.: Multi-resolution CNN for brain vessel segmentation from cerebrovascular images of intracranial aneurysm: a comparison of U-Net and DeepMedic (2020)
27. Beck, J., Rhode, S., Berkefeld, J., et al.: Size and location of ruptured and unruptured intracranial aneurysms measured by 3-dimensional rotational angiography. Surg. Neurol. **65**, 18–25 (2006)
28. Xiang, J., et al.: Hemodynamic-morphologic discriminants for intracranial aneurysm rupture. Stroke **42**, 144–152 (2011)
29. Kleinloog, R., De Mul, N., Verweij, B.H., Post, J.A., Rinkel, G.J.E., Ruigrok, Y.M.: Risk factors for intracranial aneurysm rupture: a systematic review. Neurosurgery **82**, 431–440 (2018)
30. Cebral, J.R., et al.: Analysis of hemodynamics and wall mechanics at sites of cerebral aneurysm rupture. J. NeuroInterventional Surg. **7**, 530–536 (2015)
31. Detmer, F.J.: Associations of hemodynamics, morphology, and patient characteristics with aneurysm rupture stratified by aneurysm location. Neuroradiology **61**, 275–284 (2019)
32. Detmer, F.J., et al.: Extending statistical learning for aneurysm rupture assessment to Finnish and Japanese populations using morphology, hemodynamics, and patient characteristics. Neurosurg. Focus **47**(1), E16 (2019)
33. Lindgren, A.E., et al.: Irregular shape of intracranial aneurysm indicates rupture risk irrespective of size in a population-based cohort. Stroke **47**, 1219–1226 (2016)
34. Tanioka, S., et al.: Machine learning classification of cerebral aneurysm rupture status with morphologic variables and hemodynamic parameters. Radiol.: Artif. Intell. **2**, e190077 (2020)
35. Paliwal, N., et al.: Outcome prediction of intracranial aneurysm treatment by flow diverters using machine learning. Neurosurg. Focus **45**(5), E7 (2018)
36. Suzuki, M., et al.: Classification model for cerebral aneurysm rupture prediction using medical and blood-flow-simulation data (2019)

37. Chen, G., et al.: Development and validation of machine learning prediction model based on computed tomography angiography-derived hemodynamics for rupture status of intracranial aneurysms: a Chinese multicenter study. Eur. Radiol. **30**, 5170–5182 (2020)

38. Kim, H.C., et al.: Machine learning application for rupture risk assessment in small-sized intracranial aneurysm. J. Clin. Med. **8**, 683 (2019)

39. Chandra, A.R., et al.: Initial study of the radiomics of intracranial aneurysms using Angiographic Parametric Imaging (API) to evaluate contrast flow changes (2019)

40. Silva, M.: Machine learning models can detect aneurysm rupture and identify clinical features associated with rupture. World Neurosurg. **131**, e46–e51 (2019)

41. Tachibana, Y.: A neural network model that learns differences in diagnosis strategies among radiologists has an improved area under the curve for aneurysm status classification in magnetic resonance angiography image series (2020)

42. Detmer, F.J.: Comparison of statistical learning approaches for cerebral aneurysm rupture assessment. Int. J. Comput. Assist. Radiol. Surg. **15**, 141–150 (2020)

43. Can, A., et al.: Association of intracranial aneurysm rupture with smoking duration, intensity, and cessation. Neurology **89**, 1408–1415 (2017)

44. Chabert, S., et al.: Applying machine learning and image feature extraction techniques to the problem of cerebral aneurysm rupture. Res. Ideas Outcomes **3**, e11731 (2017)

Introduction

CADA: Clinical Background and Motivation

Andreas Spuler[1] and Leonid Goubergrits[2]([⊠])

[1] Helios Hospital Berlin-Buch, Schwanebecker Chaussee 50, 13125 Berlin, Germany
andreas.spuler@helios-gesundheit.de
[2] Charité – Universitätsmedizin Berlin, Augustenburger Platz 1, 13353 Berlin, Germany
leonid.goubergrits@charite.de

Abstract. Imaging of cerebral aneurysms using DSA, MRI or CTA plays a key role in diagnosis, decision for treatment or observation, treatment planning either as microsurgical clipping or as endovascular intervention including filling of the aneurysm with coils, implantation of flow diverter in the parent vessel or filling the aneurysm with liquid embolic agents. Additionally, imaging is used in long-term follow-up of treated and untreated aneurysms. Imaging tasks and challenges include detection of aneurysms especially of aneurysms smaller than 3 mm, accurate quantitative analysis of geometric parameters assessing size and shape necessary for rupture risk assessment as well as treatment decision and eventually the type of treatment. Finally, image-based computational fluid dynamics analysis of hemodynamic risk parameters requires accurate segmentation and reconstruction of anatomical structures. These objectives motivated us to initiate the Cerebral Aneurysm Detection and Analysis (CADA) challenge. It is based on datasets of 3D rotational angiographies, the "gold standard" for clinical management of cerebral aneurysms. Datasets stem from patients with unruptured and ruptured aneurysms.

Keywords: Cerebral aneurysm · Subarachnoid bleeding · Rupture risk · Medical imaging · 3D rotational angiography

1 Background

Intracranial aneurysms (IAs) are local dilatations of the cerebral vessels, which provide the blood supply to the brain. The cerebral vessel network, subdivided into posterior and anterior circulation, is formed by two vertebral and two internal carotid arteries respectively. Both circulations are interconnected by the Circle of Willis. IAs occur at different locations in the cerebral circulation, are due to local weakening of the arterial wall, and are classified by size, shape and phenotype [1, 2]. Small aneurysms have diameters smaller than 15 mm. Larger aneurysms are subdivided into large (15 to 25 mm), giant (25 to 50 mm), and super-giant (over 50 mm) aneurysms. Small aneurysms are often subdivided into small aneurysms of less as 5 mm and middle-sized aneurysms (5–15 mm) [32]. Classified by shape, IA are subdivided into saccular and fusiform aneurysms (Fig. 1).

Prevalence of IAs varies between 1 and 5% with an average of approximately 3.2% for the adult worldwide population [3, 4]. Prevalence is influenced by a positive family

© Springer Nature Switzerland AG 2021
A. Hennemuth et al. (Eds.): CADA 2020, LNCS 12643, pp. 21–28, 2021.
https://doi.org/10.1007/978-3-030-72862-5_2

history, age, and sex, with a higher prevalence in women older than 50 [5]. The major life-threatening risk associated with aneurysms is rupture causing a subarachnoid hemorrhage (SAH). Approximately 50%-60% of all SAHs are fatal in the first month after bleeding with conservative treatment, whereas a further 25% cause morbidity of various degree. [6, 7]. Most of deaths occur during the first two weeks after bleeding. This includes 10% of deaths before hospitalization, and 25% within the first 24 h [6].

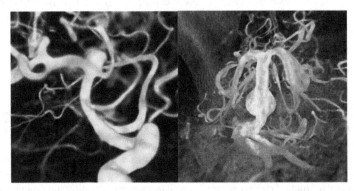

Fig. 1. Left: an example of the saccular middle cerebral artery aneurysm. Right: an example of the fusiform basilar artery aneurysm. Acquisitions were done with 3D rotational angiography. Volume rendering visualization using ZIB Amira software (Zuse Institute Berlin, Germany).

Wildely-used high-resolution medical imaging such as magnetic resonance imaging (MRI) or computed tomography (CT) results in frequent detection of unruptured intracranial aneurysms [3]. For physicians this requires the challenging decision whether to treat or to observe unruptured IAs. The challenge to balance between individual rupture risk and individual treatment risk. Rupture risk analysis is further complicated by contradicting findings: Two large studies – the International Study of Unruptured Intracranial Aneurysms (ISUIA) and the Japanese Study of Unruptured Aneurysms - found that small so far unruptured IAs almost never rupture [8, 9], whereas studies on ruptured IAs show that the majority consists of small IAs [10, 11].

2 IA Rupture Risk

Several risk factors associated with growth rate and rupture risk were identified. Altogether 24 risk factors were summarized by a systematic review [12]. Among them are site, size, specific population, arterial hypertension, sex, age, ethnicity, shape and patient history. Rupture risk increases with aneurysmal size and current clinical guidelines recommend treatment of IAs larger than 7 mm. To quantify rupture risk the PHASES score incorporating several risk factors was proposed (see Table 1) [13, 14]. PHASES calculates the 5-year risk of rupture based on pooled data from several prospective cohort studies in the USA, Canada, Netherlands, Finland and Japan. However, several studies found that PHASES might underestimate individual rupture risk thus leading to unbalanced treatment decisions [15, 16]. Aneurysm size is one controversial parameter. For the anterior circulation ISUIA found a very low rupture probability for IAs smaller than 13 mm [8]. In clinical practice, the majority of ruptured aneurysms are of small and medium size (range between 4 and 9 mm). This also applies for the cohort used in the Cerebral Aneurysm Detection and Analysis challenge. In the group ruptured aneurysms, 21% have a diameter less than 5 mm and 54% were smaller as 7 mm.

Table 1. PHASES 5-years rupture risk prediction score

Parameters	Items	Scores	Score – rupture risk
Population	• North American, European • Japanese • Finnish	0 3 5	<2 – 0.4 [0.1–1.5] 3 – 0.7 [0.2–1.5] 4 – 0.9 [0.3–2.0]
Hypertension	• no • yes	0 1	5 – 1.3 [0.8–2.4] 6 – 1.7 [1.1–2.7]
Age	• <70 years • >70 years	0 1	7 – 2.4 [1.6–3.3] 8 – 3.2 [2.3–4.4]
Size of IA (diameter)	• <7.0 mm • 7.0–9.9 mm • 10.0–19.9 mm • >20 mm	0 3 6 10	9 – 4.3 [2.9–6.1] 10 – 5.3 [3.5–8.0] 11 – 7.2 [5.0–10.2] >12 – 17.8 [15.2–20.7]
Earlier SAH from another aneurysm	• no • yes	0 1	
Site of Aneurysm	• *ICA* • *MCA* • *ACA/Pcom/posterior circulation*	0 2 4	

In addition to size, site, and sex differences, several geometric parameters of aneurysmal morphometry were associated with rupture risk, e.g. bottleneck shape, aneurysm to parent vessel size ratio, complex or irregular shape, aspect ratio (maximal aneurysm dome size to aneurysm neck size ratio). A further geometric feature affecting aneurysm rupture risk is existence of a bleb also called daughter aneurysm. Figure 2 shows some of geometric aspects associated with rupture risk.

Altogether more as 40 different geometric risk parameters were proposed [18]. Furthermore, hemodynamic parameters describing aneurysmal flow patterns as well as forces acting on aneurysmal wall such as wall shear stress or oscillatory shear index were proposed to predict aneurysm rupture. Altogether more as 15 different flow parameters were proposed [19]. Recent studies also proposed a combination of geometric and hemodynamic risk parameters to discriminate between ruptured and unruptured IAs [20–24]. Assessment of hemodynamic rupture risk parameters is based on image-based computational fluid dynamics (CFD) analysis. In turn, CFD requires accurate description of boundary conditions by segmentation and surface reconstruction of the aneurysm with parent vessel.

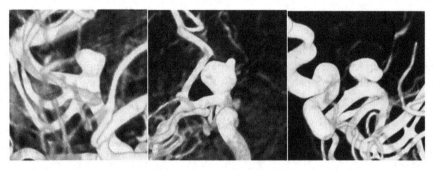

Fig. 2. Left: MCA aneurysm of an irregular saccular shape and low aspect ratio. Middle: ICA large aneurysm with a daughter aneurysm and high aspect. Right: ICA elongated aneurysm with unclear aneurysmal neck size and respectively unclear aspect ratio. Volume rendering visualization using ZIB Amira.

3   Clinical Management of IA

Two major treatment options for ruptured and unruptured IA are microsurgical clipping and endovascular occlusion [32]. Further options include hunterian ligation [33, 34] i.e. proximal ligation of the aneurysm-bearing vessel considered as the father of the micro-surgical clipping procedure, and the more recently introduced endovascular placement of flow diverting stents [35, 36], which can be considered as the third major treatment option today. Further proposed endovascular procedures include aneurysm filling with liquid embolic agents such as hardening polymers or glue [32, 37]. Coil embolization is an effective treatment procedure with a low complication rate of 3.7% but a high rate of incomplete obliteration (up to more than 40% of the cases [38, 41]). Implantation of flow diverting stents yield an occlusion rate between 62.8 and 93.9% at 1 year with a mortality rate up to 5.5% [42]. Microsurgical clipping results in very low aneurysm recurrences (1.5%) of completely clipped aneurysms, however in much more frequent recurrences in cases with incomplete clipping [39, 40, 43].

Decision for **Clipping** should take into account size, site and age [26]. Posterior circulation aneurysms are associated with a higher surgical risk compared to anterior circulation.

Decision for **Coiling** is associated with lower procedural morbidity and mortality compared to clipping but with a higher risk of recurrence [27].

Decision for **Flow Diverter** and **Liquid Embolic Agents** can be considered in carefully selected cases [26].

4   Medical Imaging of IA

Imaging plays an important role in diagnosis and treatment decision. To diagnose or observe and follow-up aneurysms respectively either noninvasive imaging modalities as computed tomography angiography (CTA) and magnetic resonance angiography (MRA) or digital subtraction angiography (DSA) are applied. Each of these modalities has advantages and disadvantages for diagnosis, follow observation or treatment procedures well described in the guidelines of the AHA/ASA (American Heart and American Stroke Associations) [26]. In spite of continuously improved imaging techniques to detect and evaluate IAs, misleading findings occur due to technical limitations associated with spatial resolution, contrast agent injection distribution, imaging artifacts or overlapping complex vessel structures [17]. Imaging should cover specific anatomic details of the aneurysm in order to select an optimal management as well as to quantify treatment outcome.

DSA is still considered the "gold standard" for IA diagnosis often as 3D rotational angiography providing a high special resolution of about 0.2 mm or less that is especially important for small aneurysms detection [25]. However, due to invasiveness and the cumulative radiation doses, DSA is less frequently used for follow-up.

CTA with a resolution of 0.4–0.7 mm is often used for IA diagnosis due to advantages of multidetector scanners. An accuracy in aneurysm detection comparable to DSA is reported for modern CT devices [28]. However, lower sensitivity was reported for aneurysms smaller as 3.0 mm [28]. However, CTA is able to support treatment decision by detection of calcifications and thrombus [28]. In ruptured IAs, CTA reliably shows SAH with a reported accuracy close to 100% [31].

MRI mainly as time-of-flight or angiography sequences has a lower resolution of 0.6–1.0 mm and shows a lower sensitivity compared with DSA or CT [31]. However, for IAs larger as 3.0 mm a high sensitivity of 89% was reported for experienced operators [29]. MRI is an effective alternative for noninvasive observation of both treated and untreated IA [30]. The use of MRI is limited in the SAH setting [31]. Long-term follow-up imaging is especially recommended for coiled aneurysms with wide aneurysmal neck, large aneurysms and aneurysms with incomplete coiling or clipping [26]. After surgical interventions imaging should assess aneurysm obliteration (complete vs. incomplete).

5 Conclusion

In summary, there are plenty motivations to use of mathematical techniques for improving image-based diagnosis, treatment decision, treatment planning, and observation of IAs including predicting rupture risk:

- **Automatic detection** of IAs with high sensitivity especially for small aneurysms of less than 3–5 mm. This could be especially important in screening of risk groups and observation of untreated as well as treated IAs. In the future, automatic detection could solve the problem of inter-observer variability especially of less experienced clinicians. Automatic aneurysm detection should be combined with automatic identification of the aneurysm site, a known rupture risk parameter.
- **Quantitative visualization** of unruptured and untreated IAs allows assessment of geometric parameters such as aneurysm and aneurysmal neck size and shape and is necessary for treatment decisions e.g. the decision between treatment or observation as well as the decision for a specific treatment option (microsurgical vs. endovascular; clipping vs hunterian ligation; coiling vs. stenting vs. a combination of stenting and coiling). Follow up of untreated aneurysms requires accurate determination of possible aneurysm grow, whereas follow-up of treated aneurysms especially in cases of incomplete coiling or clipping requires accurate assessment of possible recurrence.

- **Segmentation** and **geometry reconstruction** from imaging data is required for computational flow analysis providing hemodynamic parameters, which can be used for rupture risk analysis as well as for the selection of the optimal treatment option.

Based on these clinical motivations, CADA challenge was elaborated and limited to following three tasks: (1) aneurysm detection, (2) aneurysm segmentation, and (3) aneurysm rupture risk prediction based on provided 3D rotational angiography imaging data. Future efforts could extend the current challenge to other imaging modalities (MRI and CTA) and could include, for example, automatic identification of the aneurysm location starting with a differentiation between anterior and posterior circulation as well as further location refinement.

References

1. Keedy, A.: An overview of intracranial aneurysms. McGill J. Med. **9**, 141–146 (2006)
2. Jeong, Y.-G., Jung, Y.-T., Kim, M.-S., Eun, C.-K., Jang, S.-H.: Size and location of ruptured intracranial aneurysms. J. Korean Neurosurg. Soc. **45**, 11–15 (2009)
3. Thompson, B.G., Brown, R.D., Amin-Hanjani, S., et al.: Guidelines for the management of patients with unruptured intracranial aneurysms: a guideline for healthcare professionals from the American heart association/American stroke association. Stroke **46**, 2368–2400 (2015)
4. Brisman, J.L., Song, J.K., Newell, D.W.: Cerebral aneurysms. N. Engl. J. Med. **355**, 928–939 (2006)
5. Vlak, M.H.M., Algra, A., Brandenburg, R.J.E., Rinkel, G.J.: Prevalence of unruptured intracranial aneurysms, with emphasis on sex, age, comorbidity, country, and time period: a systematic review and meta-analysis. Lancet Neurol. **10**, 626–636 (2011)
6. Suarez, J., Tarr, R.W., Selman, W.R.: Aneurysmal subarachnoid hemorrhage. N. Engl. J. Med. **354**, 387–396 (2006)
7. Steiner, T., Juvela, S., Unterberg, A., et al.: European stroke organization guidelines for the management of intracranial aneurysms and subarachnoid haemorrhage. Cerebrovasc. Dis. **35**, 93–112 (2013)
8. Wiebers, D.O., Whisnant, J.P., Huston, J., Meissner, I., Brown, R.D., Piepgras, D.G., et al.: Unruptured intracranial aneurysms: natural history, clinical outcome, and risks of surgical and endovascular treatment. Lancet **362**, 103–110 (2003)
9. Morita, A., Kirino, T., Hashi, K., et al.: The natural course of unruptured cerebral aneurysms in a Japanese cohort. N. Engl. J. Med. **366**, 2474–2482 (2012)
10. Mokin, M., Waqas, M., Gong, A., et al.: What size cerebral aneurysms rupture? A systematic review and meta-analysis of literature. Neurosurgery **66**(1), 145–146 (2019)
11. Beck, J., Rhode, S., Berkefeld, J., et al.: Size and location of ruptured and unruptured intracranial aneurysms measured by 3-dimensional rotational angiography. Surg. Neurol. **65**, 18–25 (2006)
12. Clarke, M.: Systematic review of reviews of risk factors for intracranial aneurysms. Neuroradiology **50**, 653–664 (2008)
13. Greving, J.P., Wermer, M.J.H., Brown, R.D., et al.: Development of the PHASES score for prediction of risk of rupture of intracranial aneurysms: a pooled analysis of six prospective cohort studies. Lancet Neurol. **13**, 59–66 (2014)
14. Backes, D., Vergouwen, M.D.I., Tiel Groenestege, A.T., et al.: PHASES score for prediction of intracranial aneurysm growth. Stroke **46**, 1221–1226 (2015)

15. Bijlenga, P., Gondar, R., Schilling, S., et al.: PHASES score for the management of intracranial aneurysm: a cross-sectional population-based retrospective study. Stroke **48**, 2105–2112 (2017)
16. Darsaut, T., Fahed, R., Raymond, J.: PHASES and the natural history of unruptured aneurysms: science or pseudoscience? J Neurointerv. Surg. **9**, 527–528 (2017)
17. Howard, B.M., Hu, R., Barrow, J.W., et al.: Comprehensive review of imaging of intracranial aneurysms and angiographically negative subarachnoid hemorrhage. Neurosurg. Focus **47**, 1–3 (2019)
18. Goubergrits, L., et al.: Multiple Aneurysms AnaTomy CHallenge 2018 (MATCH) - uncertainty quantification of geometric rupture risk parameters. Biomed. Eng. Online **18**, 35 (2019)
19. Goubergrits, L., Schaller, J., Kertzscher, U., Woelken, Th., Ringelstein, M., Spuler, A.: Hemodynamic impact of cerebral aneurysm endovascular treatment devices: coils and flow diverters. Expert Rev. Med. Devices **11**, 361–373 (2014)
20. Xiang, J., Yu, J., Choi, H., Dolan Fox, J.M., et al.: Rupture resemblance score (RRS): toward risk stratification of unruptured intracranial aneurysms using hemodynamic-morphological discriminants. J Neurointerv Surg. **7**, 490–495 (2015)
21. Dhar, S., Tremmel, M., Mocco, J., et al.: Morphology parameters for intracranial aneurysm rupture risk assessment. Neurosurgery **63**, 185–196 (2008)
22. Xiang, J., Natarajan, S.K., Tremmel, M., et al.: Hemodynamic-morphologic discriminants for intracranial aneurysm rupture. Stroke **42**, 144–152 (2011)
23. Cebral, J.R., Mut, F., Weir, J., et al.: Quantitative characterization of the hemodynamic environment in ruptured and unruptured brain aneurysms. Am. J. Neuroradiol. **32**, 145–151 (2011)
24. Jou, L.D., Lee, D.H., Morsi, H., et al.: Wall shear stress on ruptured and unruptured intracranial aneurysms at the internal carotid artery. Am. J. Neuroradiol. **29**, 1761–1767 (2008)
25. Dammert, S., Krings, T., Moller-Hartmann, W., et al.: Detection of intracranial aneurysms with multislice CT: comparison with conventional angiography. Neuroradiology **46**, 427–434 (2004)
26. Thompson, B.G., et al.: Guidelines for the management of patients with unruptured intracranial aneurysms. Stroke **46**, 2368–2400 (2015)
27. Wang, H., Li, W., He, H., et al.: 320-Detector row CT angiography for detection and evaluation of intracranial aneurysms: comparison with conventional digital subtraction angiography. Clin. Radiol. **68**, e15–e20 (2013)
28. Chappell, E.T., Moure, F.C., Good, M.C.: Comparison of computed tomographic angiography with digital subtraction angiography in the diagnosis of cerebral aneurysms: a meta-analysis. Neurosurgery **52**, 624–631 (2003)
29. Li, M.H., Cheng, Y.S., Li, Y.D., et al.: Large-cohort comparison between three-dimensional time-of-flight magnetic resonance and rotational digital subtraction angiographies in intracranial aneurysm detection. Stroke **40**, 3127–3129 (2009)
30. Agid, R., Schaaf, M., Farb, R.: CE-MRA for follow-up of aneurysms post stent-assisted coiling. Interv. Neuroradiol. **18**, 275–283 (2012)
31. Hacein-Bey, L., Provenzale, J.M.: Current imaging assessment and treatment of intracranial aneurysms. AJR **196**, 32–44 (2011)
32. Lasheras, J.C.: The biomechanics of arterial aneurysms. Annu. Rev. Fluid Mech. **39**, 293–319 (2007)
33. Heros, R.C., Morcos, J.J.: Cerebrovascular surgery: past, present, and future. Neurosurgery **47**, 1007–1033 (2000)
34. Polevaya, N.V., Kalani, M.Y.S., Steinberg, G.K., et al.: The transition from hunterian ligation to intracranial aneurysm clips: a historical perspective. Neurosurg. Focus **20**, 1–7 (2006)

35. Darsaut, T.E., Bing, F., Salazkin, I., et al.: Testing flow diverters in giant fusiform aneurysms: a new experimental model can show leaks responsible for failures. AJNR Am. J. Neuroradiol. **32**, 2175–2179 (2011)
36. Augsburger, L., Reymond, P., Rufenacht, D.A., et al.: Intracranial stents being modeled as a porous medium: flow simulation in stented cerebral aneurysms. Ann. Biomed. Eng. **39**, 850–863 (2011)
37. Lanzino, G., Kanaan, Y., Perrini, P., et al.: Emerging concepts in the treatment of intracranial aneurysms: stents, coated coils, and liquid embolic agents. Neurosurgery **57**, 449–459 (2005)
38. Brilstra, E.H., Rinkel, G.J., van der Graaf, Y., et al.: Treatment of intracranial aneurysms by embolization with coils: a systematic review. Stroke **30**, 470–476 (1999)
39. David, C.A., Vishteh, A.G., Spetzler, R.F., et al.: Late angiographic follow-up review of surgically treated aneurysms. J. Neurosurg. **91**, 396–401 (1999)
40. Brown, M.A., Parish, J., Guandique, C.F., et al.: A long-term study of durability and risk factors for aneurysm recurrence after microsurgical clip ligation. J. Neurosurg. **126**, 819–824 (2017)
41. Colby, G.P., Paul, A.R., Radvany, M.G., et al.: A single center comparison of coiling versus stent assisted coiling in 90 consecutive paraophthalmic region aneurysm. J. Neurointerv. Surg. **4**, 116–120 (2012)
42. Campos, J.K., Cheaney Ii, B., Lien, B.V., et al.: Advances in endovascular aneurysm management: flow modulation techniques with braided mesh devices. Stroke Vasc. Neurol. **5**, 1–3 (2020)
43. Burkhardt, J.-K., Chua, M.H.J., Miriam Weiss, M., et al.: Risk of aneurysm residual regrowth, recurrence, and de novo aneurysm formation after microsurgical clip occlusion based on follow-up with catheter angiography. World Neurosurg. **106**, 74–84 (2017)

Cerebral Aneurysm Detection

Deep Learning-Based 3D U-Net Cerebral Aneurysm Detection

Matthias Ivantsits[1(✉)], Jan-Martin Kuhnigk[2], Markus Huellebrand[1,2], Titus Kuehne[1,3], and Anja Hennemuth[1,2,3]

[1] Charité – Universitätsmedizin Berlin, Augustenburger Pl. 1, 13353 Berlin, Germany
`matthias.ivantsits@charite.de`
[2] Fraunhofer MEVIS, Am Fallturm 1, 28359 Bremen, Germany
[3] German Heart Institute Berlin, Augustenburger Pl. 1, 13353 Berlin, Germany

Abstract. Subarachnoid hemorrhage, commonly caused by the rupture of cerebral aneurysms, is a life-threatening condition with high mortality and morbidity. With a death rate of roughly 40%, it is highly desirable to detect aneurysms early and decide about the appropriate rupture prevention strategy. Rotational X-ray angiography is a non-invasive imaging modality and enables diagnostics to detect cerebral aneurysms at an early stage.

We propose a variation of the 3D U-Net architecture for the detection and localization of these cerebral aneurysms. This model is enhanced with a knowledge-based postprocessing strategy to minimize the false-positive detections per case. Our suggested method shows similar sensitivity statistics compared to state-of-the-art solutions, with a drastically reduced false-positive rate per patient. The described solution is almost entirely accurate on structures larger than 5 mm in diameter but shows difficulties with smaller aneurysms. We show an F2-score of 0.84 and a false-positive rate of 0.41 on a private test set.

Keywords: Cerebral aneurysms · Subarachnoid hemorrhage · X-ray rotational angiography · Deep learning · Machine learning · Automatic detection

1 Introduction

Subarachnoid hemorrhage (SAH) caused by an aneurysm rupture is a life-threatening condition with high mortality and morbidity. The death rate is above 40% [2], and in case of survival, cognitive impairment can affect patients for a long time, even lifelong. It is therefore highly desirable to detect aneurysms early and decide about the appropriate rupture prevention strategy. Rotational X-ray angiography enables imaging diagnostics and can be utilized to detect cerebral aneurysms at an early stage. Figure 1 illustrates two exemplary images with highlighted aneurysms. The proposed method of this work is used to participate in the "Cranial aneurysm detection and analysis challenge"[1] (CADA), hosted by Charité Berlin, BIFOLD, Fraunhofer MEVIS, and Helios.

---

[1] https://cada.grand-challenge.org/.

© Springer Nature Switzerland AG 2021
A. Hennemuth et al. (Eds.): CADA 2020, LNCS 12643, pp. 31–38, 2021.
https://doi.org/10.1007/978-3-030-72862-5_3

Fig. 1. Two exemplary rotational X-ray angiographic images with highlighted cerebral aneurysms.

Conventional digital image processing methods are based on hand-engineered features to detect structures in images. These features are comprised of different convolutional kernels, gray-level thresholds, vessel curvature, and other geometric components. Methods based on these conventional image processing approaches [6–13] illustrate acceptable performance metrics. From 2018 on deep learning (DL) seems to be the gold standard for the detection and localization of cerebral aneurysms [4,14–22]. More sophisticated and specialized DL networks [3,23–28] present solutions with slight improvements in accuracy and false-positive rates. Moreover, the inference time is drastically reduced by applying these architectures.

Most available methods in the literature are either DL-based or based on classical image processing methods. We want to ease this distinction by applying a standard 3D U-Net extended by a knowledge-based postprocessing strategy to reduce the number of false-positive detections per case.

2   Method and Materials

2.1   Dataset

In total, the dataset consists of 131 rotational X-ray angiographic images from different patients. A private dataset of 22 images was held back as a test set by the CADA challenge organizers during the training period, and only later released to allow for offline processing and to send in the results for centralized evaluation. The remaining 109 cases were available for model training and validation.

2.2   Architecture

For the localization of cerebral aneurysms, we use a variation of the 3D U-Net architecture introduced by Ronneberger et al. [29], which has shown to be successful in various domains, especially in medical image segmentation. The original model was developed for biomedical image segmentation, more specifically to segment cells. The architecture has a standard encoder-decoder block. The architecture is purposely designed not to contain any fully connected (FC) layers. The feature maps only contain pixels for which the full context in the input image is available.

There are some architectural changes we have introduced to the standard U-Net model. First, we reduced the encoder and decoder to three levels, due to GPU memory constraints. Next, we avoid bottlenecks, as proposed by Szegedy et al. [30], by doubling the number of channels before max pooling. Moreover, we introduce a batch normalization, which was introduced by Ioffe et al. [31]. Batch normalization is a regularization method and counteracts the internal covariate shift of the network's parameters. This makes the network easier to train with higher learning rates, less likely to overfit, and less sensitive to the initialization of the parameters. Furthermore, we utilize spatial dropout, introduced by Tompson et al. [32]. This is another technique to improve a models generalization capability, by dropping out entire feature maps from a convolutional layer and preventing activations from becoming strongly correlated.

Preprocessing steps we have performed are applied during model training as well as inference time. First, we apply a Lanczos resampling of the image to a voxel size of 1 mm × 1 mm × 1 mm, which results in images of size 72 × 72 × 60. Next, we normalize the image by a percentile mapping. The 1- and 99-percentile of the image intensities are mapped to 0 and 1, respectively. The linear scaling is applied without clipping at the boundaries. Due to the small volume in relation to the background, the intensity of vessels and aneurysms is typically located above the 99-percentile, resulting in normalized values greater than 1. Additionally, we enhance this image with an additional channel containing a vessel segmentation. We apply a threshold ≥ 1.5 followed by a morphological closing operation with a 6-neighborhood. Moreover, we calculate for each foreground voxel the euclidean distance to the closest background voxel. This results in high values for the center of thick vessels and aneurysms. Concatenating the normalized image with this additional vessel segmentation produces an image of size H × W × D × 2. Finally, we divide the image into 28 × 28 × 28 patches for training due to memory constraints. For the border, we use a reflection padding of 20 × 20 × 20.

Model training was performed utilizing Keras with the TensorFlow backend. Training employs a batch size of ten, with eight images containing at least one foreground voxel and the remaining two enclosing only background voxels. Furthermore, we utilize data augmentation during training. First, we apply an orthogonal patch flipping with a probability of 0.5 in each dimension (x, y, z). Moreover, we add some random noise with a standard deviation of 0.2 and a Gaussian smoothing with sigma in the range of 0.1 to 0.3. Finally, an intensity shift with a standard deviation of 0.2 and a scaling of the image intensities between 0.5 and 1.5 is applied. For model optimization, we use Adam with standard parameters and a learning rate of 0.001 and train for 30,000 iterations. We use a categorical cross-entropy loss with per-voxel weights that make use of the ground truth. By taking both image intensity (which is high at the vessel/aneurysm centers) and the true aneurysm locations into account, the weights were designed to optimize sensitivity in this detection task. Aneurysm (center) voxels received maximum weights, decreasing to their periphery. Other high intensity (mostly vessel) voxels that are prone to produce false positives received medium weights, background voxels the lowest. The model was trained on 5 different splits to reduce the risk of lousy performance due to an unfortunate split. The 109 training cases were split into 79 cases for training, 9 for validation, and 21 for model testing.

Postprocessing of the results produced by the U-Net includes several steps. First, we binarize the prediction results at a threshold of 0.5, followed by connected component analysis proposed by Ng et al. [33] to separate the resulting mask into individual aneurysm candidates. This analysis is performed on an 18-neighborhood. Next, we remove all candidates that are not connected to any vessel segments calculated in the preprocessing step of the second channel. This step is followed by the removal of all candidates that touch the image border. Experiments provided evidence that despite the reflection padding, several false-positives were detected, while no true positives touched the border. After that, we apply a trilinear upsampling to the original resolution with a binarization at a threshold of 0.5. Finally, we calculate the point within the candidate mask furthest away from the structure surface and use it as the localization vector for the detection.

3 Experiments and Results

3.1 Execution Hardware and Computation Time

The presented experiments were conducted using an Nvidia RTX 2080 Ti GPU with 10 GB memory and Intel Core i9 9820X @ 3.30 GHz CPU. On this hardware, the total average processing time per image was 2.25 ± 0.04 s. The image preprocessing took 0.21 s, and the postprocessing steps 0.19 on average, the GPU-based inference used the remaining time.

3.2 Best Model Selection and Performance Evaluation

We selected the best model from the 5-split validation based on the F2-score, where the F2-score combines sensitivity and precision and considers sensitivity twice as important as precision. Table 1 shows the 5-split validation metrics containing the F2-score, sensitivity, precision, F2-score, the number of false-positives per case, dice-score, and the Jaccard-score, each computed on their respective test sets. According to these metrics we chose the model from split 5 to be optimal and continue to apply it to the private test set.

Table 1. The 5-split validation metrics containing the F2-score, sensitivity, precision, F2-score, the number of false-positives per case, dice-score, and the Jaccard-score, each computed on their respective test sets.

Split	Sensitivity	Precision	F2-score	#FP/case	Dice	Jaccard
1	80.77	77.78	0.80	0.27	0.84	0.74
2	84.61	73.33	0.82	0.36	0.84	0.75
3	80.00	83.33	0.81	0.18	0.86	0.77
4	88.89	60.00	0.81	0.73	0.80	0.70
5	100.00	71.86	0.93	0.43	0.87	0.78
Avg.	**86.85**	**73.26**	**0.83**	**0.39**	**0.84**	**0.75**

On the training set with 109 cases, the selected model produced 91 false-positives without any postprocessing. Applying the filter removing candidates without connections to the vessels reduced the false-positives to 43 and the image border filter further to 36. This results in 0.33 false-positives on average per case.

Lastly, we want to emphasize the detection statistics by aneurysm size. The presented results were compiled from all five trained models from the split, each evaluated on its test set. Figure 2 highlights these detection statistics, where small aneurysms are $<5\,\mathrm{mm}$, medium between $5\,\mathrm{mm}$ and $15\,\mathrm{mm}$, and large structures $\geq 15\,\mathrm{mm}$. The sensitivity on large vessel dilations results in 100%, for medium-sized in 100%, and small-sized in 65%.

3.3 Performance on the CADA Challenge Test Set

The private test set of the CADA challenge consists of 22 images with 30 cerebral aneurysms. The evaluation performed by the challenge organization revealed that the proposed method achieved a sensitivity of 86.6% with 0.41 false-positives per case, resulting in an F2-score of 0.84.

4   Discussion

We have illustrated a process with similar sensitivity statistics mentioned in the literature and a very low false-positive detection rate compared to state-of-the-art solutions. Especially our postprocessing has shown to substantially decrease the number of false-positives per case, which makes the architecture more applicable in clinical settings and less frustrating for physicians to use.

While the proposed method proved to be very sensitive for all medium- and large-sized aneurysms in the performance evaluation experiments, there remain problems with small aneurysms (smaller than $5\,\mathrm{mm}$), as highlighted in Fig. 2. The literature has shown great difficulties with these delicate structures. One potential solution could be increasing the input resolution of the images. By decreasing the voxel size, more minuscule structures will be available to the

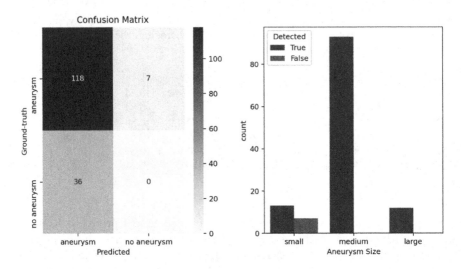

Fig. 2. The confusion matrix on the 5-split test sets displayed on the left side. The right chart illustrates the true-positives and false-negatives classified by aneurysm size, where small aneurysms are <5 mm, medium between 5 mm and 15 mm, and large structures ≥15 mm.

model during training and inference. Furthermore, more substantial data augmentation, like an elastic deformation likely increases the overall models' performance. The accuracy could be further enhanced by applying transfer learning with weights learned on other datasets.

Another probable advancement of the proposed method is forcing a separation of detected structures. We discovered issues with aneurysms that are very close or even touching to be detected as single structure during our experiments. Improving the separation of the aneurysms is especially relevant for methods that built on top of an automatic detections, e.g. rupture risk estimations. This false detection can be improved with an idea proposed by Ronneberger et al. [29], by introducing a loss that is forcing the network to learn border voxels.

5   Conclusion

We have proposed a novel combination of DL and conventional image processing to detect and localize cerebral aneurysms which showed a sensitivity comparable to state-of-the-art solutions [15, 20, 25, 28, 34] and a substantially decreased false-positive rate. The method can be utilized in automatic aneurysm screening workflows, and furthermore, used as a preceding step for rupture risk assessments or treatment strategies to an attending physician.

References

1. Bhidayasiri, R., et al.: Neurological differential diagnosis: a prioritized approach (2005)
2. Teunissen, L.L., et al.: Risk factors for subarachnoid hemorrhage (1996)
3. Park, A., et al.: Deep learning-assisted diagnosis of cerebral aneurysms using the HeadXNet model. JAMA Netw. Open **2**, e195600–e195600 (2019)
4. Faron, A., et al.: Performance of a deep-learning neural network to detect intracranial aneurysms from 3D TOF-MRA compared to human readers. Clin. Neuroradiol. **30**, 591–598 (2019)
5. Hirai, T., et al.: Intracranial aneurysms at MR angiography: effect of computer-aided diagnosis on radiologists' detection performance. Radiology **237**, 605–610 (2005)
6. Arimura, H., et al.: Automated computerized scheme for detection of unruptured intracranial aneurysms in three-dimensional magnetic resonance angiography. Acad. Radiol. **11**, 1093–1104 (2004)
7. Lauric, A., et al.: Automated detection of intracranial aneurysms based on parent vessel 3D analysis. Med. Image Anal. **14**(2), 149–159 (2010)
8. Yang, X., et al.: Computer-aided detection of intracranial aneurysms in MR angiography. J. Digit. Imaging **24**(1), 86–95 (2011)
9. Hentschke, C., et al.: Detection of cerebral aneurysms in MRA, CTA and 3D-RA data sets (2012)
10. Hentschke, C., et al.: A new feature for automatic aneurysm detection (2012)
11. Chen, S.-P., et al.: Evaluation of imaging diagnosis and assessment value of three-dimensional digital angiography for intracranial aneurysms (2012)
12. Koc, K., et al.: Detection and evaluation of intracranial aneurysms with 3D-CT angiography and compatibility of simulation view with surgical observation (2014)
13. Sulayman, N., et al.: Semi-automatic detection and segmentation algorithm of saccular aneurysms in 2D cerebral DSA images. Egypt. J. Radiol. Nucl. Med. **47**(3), 859–865 (2016)
14. Nakao, T., et al.: Deep neural network-based computer-assisted detection of cerebral aneurysms in MR angiography. J. Magn. Reson. Imaging **47**(4), 948–953 (2018)
15. Rahmany, I., et al.: A fully automatic based deep learning approach for aneurysm detection in DSA images (2018)
16. Ueda, D., et al.: Deep learning for MR angiography: automated detection of cerebral aneurysms. Radiology **290**(1), 187–194 (2019)
17. Joo, B., et al.: A deep learning algorithm may automate intracranial aneurysm detection on MR angiography with high diagnostic performance. Eur. Radiol. **30**, 5785–5793 (2020)
18. Stember, J., et al.: Convolutional neural networks for the detection and measurement of cerebral aneurysms on magnetic resonance angiography. J. Digit. Imaging **32**(5), 808–815 (2019)
19. Chen, G., et al.: Automated computer-assisted detection system for cerebral aneurysms in time-of-flight magnetic resonance angiography using fully convolutional network. BioMed. Eng. OnLine **19**, 1–10 (2020)
20. Jin, H., et al.: Fully automated intracranial aneurysm detection and segmentation from digital subtraction angiography series using an end-to-end spatiotemporal deep neural network. J. NeuroInterventional Surg. **12**(10), 1023–1027 (2020)

21. Sichtermann, T., et al.: Deep learning-based detection of intracranial aneurysms in 3D TOF-MRA. Am. J. Neuroradiol. **40**(1), 25–32 (2019)
22. Patel, T., et al.: Multi-resolution CNN for brain vessel segmentation from cerebrovascular images of intracranial aneurysm: a comparison of U-Net and DeepMedic (2020)
23. Zhang, Y., et al.: DDNet: a novel network for cerebral artery segmentation from MRA images (2019)
24. Dai, X., et al.: Deep learning for automated cerebral aneurysm detection on computed tomography images. Int. J. Comput. Assist. Radiol. Surg. **15**, 715–723 (2020)
25. Zeng, Y., et al.: Automatic diagnosis based on spatial information fusion feature for intracranial aneurysm. IEEE Trans. Med. Imaging **39**(5), 1448–1458 (2020)
26. Zhou, M., Wang, X., Wu, Z., Pozo, J.M., Frangi, A.F.: Intracranial aneurysm detection from 3D vascular mesh models with ensemble deep learning. In: Shen, D., et al. (eds.) MICCAI 2019. LNCS, vol. 11767, pp. 243–252. Springer, Cham (2019). https://doi.org/10.1007/978-3-030-32251-9_27
27. Yang, X., et al.: Surface-based 3D deep learning framework for segmentation of intracranial aneurysms from TOF-MRA images (2020)
28. Duan, H., et al.: Automatic detection on intracranial aneurysm from digital subtraction angiography with cascade convolutional neural networks. Biomed. Eng. Online **18**(1), 1–18 (2019)
29. Ronneberger, O., Fischer, P., Brox, T.: U-Net: convolutional networks for biomedical image segmentation. In: Navab, N., Hornegger, J., Wells, W.M., Frangi, A.F. (eds.) MICCAI 2015. LNCS, vol. 9351, pp. 234–241. Springer, Cham (2015). https://doi.org/10.1007/978-3-319-24574-4_28
30. Szegedy, C., et al.: Rethinking the inception architecture for computer vision (2015)
31. Ioffe, S., et al.: Batch normalization: accelerating deep network training by reducing internal covariate shift (2015)
32. Tompson, J., et al.: Efficient object localization using convolutional networks (2015)
33. Ng, A.Y., et al.: On spectral clustering: analysis and an algorithm. Adv. Neural Inf. Process. Syst. **2**, 849–856 (2001)
34. Sulayman, N., et al.: Semi-automatic detection and segmentation algorithm of saccular aneurysms in 2D cerebral DSA images. Egypt. J. Radiol. Nuclear Med. **47**(3), 859–865 (2016)

Detect and Identify Aneurysms Based on Adjusted 3D Attention UNet

Yizhuan Jia[2], Weibin Liao[1], Yi Lv[1], Ziyu Su[1], Jiaqi Dou[3], Zhongwei Sun[2(✉)], and Xuesong Li[1(✉)]

[1] School of Computer Science and Technology, Beijing Institute of Technology, Beijing, China
lixuesong@bit.edu.cn
[2] Mediclouds Medical Technology, Beijing, China
szw@mediclouds.cn
[3] Center for Biomedical Imaging Research, Department of Biomedical Engineering, School of Medicine, Tsinghua University, Beijing, China

Abstract. Early diagnosis and treatment of cerebral aneurysms are important for reducing the risk of aneurysm rupture. Fast and accurate detection of aneurysms on blood vessels is a key step in diagnosis of aneurysm. To date, a large number of deep learning algorithms, especially the UNet network, have been developed for detection of aneurysms. However, when the amount of data for training is small, it is difficult to obtain a reliable deep learning network to effectively identify aneurysms. In order to address this issue and improve the accuracy of aneurysm detection, here we proposed to combine the deep learning approach with specially designed preprocessing and postprocessing algorithm. We first determined the rough locations of the aneurysms based on the features on the vascular skeleton before aneurysms segmentation with deep learning network, i.e. 3D Attention UNet in this work, thus reducing the missed detection rate of the UNet network. We could obtain the shape and texture related to the aneurysm. Then we used the random forest algorithm to implement the feature classification model to find out the false aneurysms incorrectly detected by the U-Net network. The experimental results show that our method can accurately identify aneurysms in the case of small data sets.

Keywords: Aneurysm detection · Random forest · 3D attention UNet

1 Introduction

Aneurysm is characterized by local structural deterioration of the arterial wall, with loss of the internal elastic lamina and disruption of the media [1]. Patients with unruptured cerebral aneurysm usually do not show any symptom, but the rupture of the cerebral aneurysm will cause subarachnoid hemorrhage, leading to death or disability and other consequences, seriously threatening the life of the patient. Therefore, early diagnosis and treatment of cerebral aneurysms are important for reducing the risk of aneurysm

---

Y. Jia and W. Liao—These authors contributed equally to this work.

© Springer Nature Switzerland AG 2021
A. Hennemuth et al. (Eds.): CADA 2020, LNCS 12643, pp. 39–48, 2021.
https://doi.org/10.1007/978-3-030-72862-5_4

rupture. At the moment, Digital Subtraction Angiography (DSA) is the gold standard imaging technique for diagnosing aneurysms, with a sensitivity of over 95% [2]. DSA images can accurately reflect the location, scope, extent, and branch of the disease.

To date, a large number of effective algorithms and models have been proposed for detection of cerebral aneurysm.

In recent years, several studies have applied deep learning and convolutional neural networks to aneurysms detection. Takahiro Nakao and Shouhei Hanaoka [3] developed an automatic intracranial aneurysm detection system based on non-enhanced magnetic resonance angiography images with the deep convolutional neural networks and maximum intensity projection algorithms. The detection model trained by this system is highly sensitive and has a relatively low false positive rate. In the detection of coronary artery disease, Jelmer M. Wolterink [4] and others proposed an algorithm that used a convolutional neural network to extract the centerline of the coronary artery in cardiac CT angiography (CCTA) image. They used the trained 3D dilated convolutional neural network to predict the most likely direction and radius of the artery at any given point in the CCTA image to obtain the centerline of the blood vessel and other features for detecting coronary atherosclerosis and coronary aneurysms. This method also had important implications for the detection of intracranial aneurysms. Daiju Ueda and Akira Yamamoto [5] adopted ResNet-18 and used hundreds of arterial anomalies on 3D TOF MR angiography images for training. The obtained network can detect aneurysms in the image, and the sensitivity of the internal and external test data sets is high.

Dataset: The training dataset released by the organizers of the CADA Challenge included 109 image data, 127 aneurysms with coordinates annotation. The test cases for generating final results contained 22 image data.

Our Contributions: In this work, we used the 3D Attention UNet network to obtain the edge of aneurysm. However, the result by the network is not perfect. Therefore, we proposed to combine the 3D Attention UNet network with the characteristic information of the aneurysm on the skeleton to ensure that the model will not miss the aneurysm. Besides, we performed further classification of the aneurysms that were segmented by the UNet based on the morphological and graphic features of the aneurysms, thus reducing the false positive rate.

2   Method

We first adopted a highly sensitive method to capture the aneurysms in the blood vessel as far as possible, regardless the high false positive rate. In the second step, an aneurysm filtering strategy is designed to screen out falsely detected aneurysms. In the third step, in order to obtain more features that are different between false positive aneurysms and real aneurysms, and also to obtain detailed outline features of aneurysms, we used the 3D Attention UNet network to segment the aneurysm.

2.1 Excessive Detection of Aneurysm Points

Thresholding segmentation was performed to the original images, with the threshold value manually selected such that majority of the blood vessels were kept in the segmented images while noise level is acceptable. We indicated the point set on the segmented image as V, and used the skeleton extraction algorithm to obtain the skeleton of the blood vessel. Then we indicated the point set on the skeleton as S, and obtained the possible center point of the aneurysm by calculating and searching the features on the blood vessel skeleton line.

The contours of the blood vessels, indicated as C, were obtained by eroding the segmented images and performing XOR operation between the eroded segmented images and the original segmented images. Based on the contour, we can obtain the contour curve associated with any point on the center line for feature calculation (see Fig. 1 (a)).

$$C = V \oplus V'$$ (1)

The point c on the blood vessel contour C was associated with the point s on the skeleton line S (see Fig. 1 (b)) according to the following condition

$$\forall s' distance(c, s) \leq distance(c, s')$$ (2)

where $c \in C$, $s, s' \in S$。

Then we can bind a contour curve Cr to any point on the skeleton, and calculate the characteristic value of the point on the skeleton line based on this curve.

We selected three features with high sensitivity to aneurysms to search. For the point s on the skeleton line S, we calculated the following three features:

numC: the number of points on the contour curve Cr associated with s.
maxR: the distance between the point farthest from s and c on the contour curve associated with s

$$maxR = distance(c, s), and\ \forall c', distance(c, s) \geq distance(c', s)$$ (3)

where $c, c' \in Cr$.
avgR: the average distance between all points on the contour curve and the center point

$$avgR = \frac{1}{numC} \times \sum_{c \in Cr} distance(c, s)$$ (4)

As we know, under normal circumstances, the radius of the blood vessel where there is an aneurysm tends to be relatively large, so the feature *maxR* and *avgR* can help us locate the location of the aneurysm. However, for aneurysms at vascular bifurcations, such radius-based detection is often less effective. Therefore, we added *numC* as a new feature to improve the detection. According to our experiments, we found that in the blood vessel where the aneurysm is located, the value of *numC* would increase significantly.

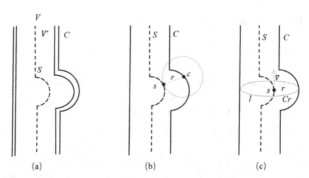

Fig. 1. (a) Demonstrates calculation of the contour of the blood vessel. (b) demonstrates calculation of the points on the skeleton line and the points on the vessel contour. (c) demonstrates calculation of the features

In order to further strengthen the generalization ability of point s for aneurysm features, we calculated the gradient of the above features along the skeleton line and used them as new features to further narrow the search range of aneurysm points (see Fig. 1 (c)).

We empirically specified the ranges of the above characteristics of the center point for identifying aneurysm, by checking the characteristic values for the aneurysms in the existing data sets. In fact, using the specified ranges, our model can detect all aneurysm points in the blood vessel. This is because aneurysm points we detected exist on the center line of the blood vessel. Although this may cause them not be inside aneurysms, we can ensure that there are aneurysm points detected near each aneurysm. On the other hand, it should be noted that this strategy inevitably leads to many false positive aneurysm points.

2.2 Segmentation of Aneurysm

The next step is to remove the false positive points from the above-detected candidate points as far as possible. To achieve this, more aneurysm characteristics need to be calculated and used. In this study, this was done by using pretrained 3D Attention UNet network to perform segmentation of the aneurysm (Fig. 2). We constructed the 3D UNet with attention gates, with the initial number of convolution kernels set to 44 and convolutional kernel sizes to $3 \times 3 \times 3$. The 3D Attention UNet network employs the attention mechanism to supervise the feature concatenation from shallow layer features to deeper, which suppresses the concentration on background and enhance the vessel regions, thus improving the segmentation performance and achieving end-to-end segmentation.

In the segmentation stage, we iterated through every seed point to perform segmentation and labeling on the center of input image cube simultaneously until all seed points are labeled. In addition, we apply Opening (morphology) on the segmentation mask to separate two adjacent aneurysms after segmentation.

Pre-training Stage: In order to reduce the impact on the network learning ability due to the small amount of data, and to ensure that the network can learn the three-dimensional

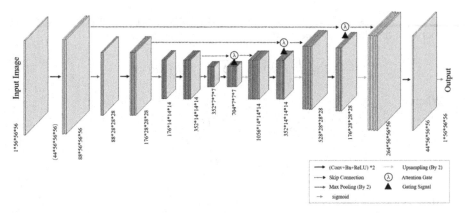

Fig. 2. The structure of the 3D Attention U-Net

information of the image, we used transfer learning model to pre-train the deep learning network. We used a self-supervised learning method called Models Genesis [6] for pre-training. The general pre-training strategy uses Autoencoder to let the Encoder and Decoder learn features through the process of image reconstruction, and then use the trained Encoder and Decoder to do segmentation or classification tasks. The Encoder of Models Genesis does not use the original image as input. Instead, images obtained from three types of noise generating methods (non-linear transformation, local pixel shuffling, out-painting and in-painting), were used as the input. The transformed image is restored to the original image through the Encoder and Decoder, so that the network can learn the shape features, texture features and contextual characteristics.

2.3 Screening for False Positive Aneurysms

Although we have used the UNet network to initially screen the detected aneurysms, there are still some false positive aneurysms. Therefore, further refinement of the result is needed. To achieve this, Radiomic [7] was used to extract additional features from the aneurysm body extracted by the UNet network. Then random forest model was used to perform classification of these additional features to identify the true aneurysms.

2.4 Evaluation Model

We used Mean Average Precision (MAP) to evaluate the performance of our detection model. The MAP was proposed in the Pascal Visual Objects Classes challenge, and has been shown to be useful for predicting the location and category of objects.

To evaluate the accuracy of detection, a metric for the confidence of the predicted bounding box is needed. For the MAP, the metric we used is Intersection over Union (IoU). This is a very simple visual quantity, and its calculation formula is as follows:

$$IoU = \frac{Prediction \cap Ground\ Truth}{Prediction \cup Ground\ Truth} \tag{5}$$

In the VOC Challenge, the calculation of Average Precision (AP) was defined as the area between the Precision-Recall (PR) curve and the coordinate axis:

$$AP = \int_0^1 P(R)dR = \sum_{k=0}^n P(k)\Delta R(k) \tag{6}$$

Where,

$$Precision = \frac{TP}{TP + FP} \tag{7}$$

$$Recall = \frac{TP}{TP + FN} \tag{8}$$

MAP is the average of all APs in all categories.

3   Experiments and Results

The experiments were performed on the 109 cases of data provided by the competition. According to the selected features on the skeleton line, we could finally capture 3549 suspicious aneurysm points, of which the number of points verified as true positive aneurysms was 917 and the number of false positive aneurysm points was 2632. It can be seen that such a search method would detect multiple suspicious aneurysm points on the same aneurysm, and there are many false positive aneurysm points (see Fig. 3).

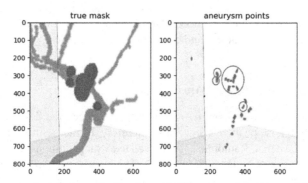

Fig. 3. The figure on the left shows the blood vessel (light red part) and the actual aneurysm (dark red part). The figure on the right shows the detected aneurysm points (red part), but only points in these blue circle parts are the true positive aneurysm points (Color figure online).

Next, we used 3D Attention UNet to segment the aneurysm. In the training step, we set the batch size to 4, and trained the model for 100 epochs, and the shape and spacing of input image cube is $56 \times 56 \times 56$ and $0.5 \times 0.5 \times 0.5$, respectively. The training process will stop automatically when validation loss doesn't decrease for 15 epochs. Adam optimizer was used to update the weights of the network with an initial learning rate of $\alpha = 10^{-4}$, six-fold cross validation was employed, in order to get a reliable model.

In this step, we can achieve the following functions:

(1) Preliminary screening of the false positive aneurysm points detected in the previous step.
(2) Combined processing of multiple aneurysm points detected on the same true positive aneurysm
(3) Obtaining complete aneurysms

In the experiment, we obtained a total of 211 aneurysm data after UNet processing. Among the obtained aneurysms, 125 cases were true aneurysms and 86 were false aneurysms.

For these 211 aneurysms, we used Radiomics to extract features from their 3-dimensional data. The extracted features included first-order statistics, three-dimensional size and shape descriptors, Gray Level Co-occurence Matrix (GLCM), Gray Level Run Length Matrix (GLRLM), Gray Level Size Zone Matrix (GLSZM), Neighborhood Gray-Tone Difference Matrix (NGTDM), Gray Level Dependence Matrix (GLDM). Thus, a total of 906 valid features were extracted.

We labeled these aneurysms, divided the data according to a 4:1 ratio, and used the random forest classification model for training and prediction.

The methods involved are as follows:

(1) Use GridSearchCV to adjust model parameters
(2) Use a 5-fold cross validation method.
(3) Use the feature selection function of random forest to identify the key features to achieve data dimensionality reduction, and reduce the complexity of the model without sacrificing the performance of the model.
(4) Use ROC_AUC as the evaluation standard of the classification model, and select the most ideal ROC curve while minimizing missed and false detections.

The experimental results show that the AUC value of our model is 0.9756 under the most ideal ROC curve (see Fig. 4).

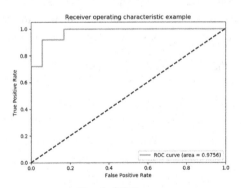

Fig. 4. ROC curve

In the case of selecting 0.5 as the classification threshold, the classification results of 43 test examples (20 false positive aneurysms, 23 true positive aneurysms) are as shown in Table 1.

Table 1. Random forest - Confusion matrix

Real predit	False	True
False	18	2
True	0	23

It is understandable that the detection model has higher performance for true positive aneurysms, but it is still possible to result in some false aneurysms. In order to further make a reasonable evaluation of the model, we drew the PR curve of the detection model and calculated its MAP value.

When 0.8 was selected as the confidence threshold of IoU, the following PR diagram was obtained, (see Fig. 5) with the MAP value being 0.9711.

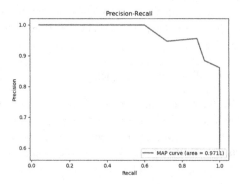

Fig. 5. Precision-Recall curve

The intermediate result of a typical sample is shown in Table 1 and Fig. 6. In this example, there is only one true positive aneurysm, but we can find several aneurysm points using the feature method extracted on the skeleton line, and after the use of 3D Attention UNet, there still remain false aneurysms. But the random forest algorithm can find out and remove all these false aneurysms (Table 2).

Table 2. Report of accuracy.

Method	Precision	Recall
aneurysm points	917/3549 = 0.26	1.0
3D Attention UNet	125/211 = 0.59	125/128 = 0.97
Random forest	123/131 = 0.94	123/125 = 0.98
Overall	0.94	0.96

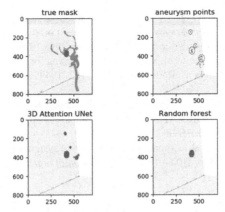

Fig. 6. The upper left panel shows the true aneurysm, and the upper right panel second figure shows these aneurysm points extracted on the skeleton line (only the points in the blue circle are true aneurysm points, and the points in the green circle are false positive aneurysm points that 3D Attention Unet cannot identify), and the bottom left panel shows aneurysms obtained by segmentation using the 3D Attention UNet, and the bottom right panel shows the aneurysm obtained after screening using the random forest algorithm (Color figure online)

4 Conclusion

This paper mainly introduces an optimization method for aneurysm detection based on the 3D Attention UNet network with specially designed preprocessing and postprocessing strategies. We use the characteristic information of the aneurysm on the skeleton and the 3D voxel information of the aneurysm itself, to refine the detection. Experiments show that at this stage our model is extremely sensitive to true positive aneurysms and the specially designed refining strategies were effective for improving the detection accuracy. Considering that it is difficult to obtain a large amount of data to train a good deep learning model, the method proposed in this study may have great potential for use in clinical practice.

From the experimental results, the model still detects some false-positive aneurysms. The characteristics of these aneurysms are similar to real aneurysms, so that our model cannot identify them well. Studying these false positive aneurysms and discovering their unique characteristics to effectively distinguish them is the focus of this research in the future.

References

1. Drake, C.: Management of cerebral aneurysm. Stroke **12**(3), 273–283 (1981)
2. Kato, Y., Katada, K., Hayakawa, M., et al.: Can 3D-CTA surpass DSA in diagnosis of cerebral aneurysm? Acta Neurochir. **143**(3), 245–250 (2001). https://doi.org/10.1007/s007010170104
3. Nakao, T., Hanaoka, S., Nomura, Y., et al.: Deep neural network-based computer-assisted detection of cerebral aneurysms in MR angiography. J. Magn. Reson. Imaging **47**(4), 948–953 (2018). https://doi.org/10.1002/jmri.25842

4. Wolterink, J.M., van Hamersvelt, R.W., Viergever, M.A., et al.: Coronary artery centerline extraction in cardiac CT angiography using a CNN-based orientation classifier. Med. Image Anal. **51**, 46–60 (2019). https://doi.org/10.1016/j.media.2018.10.005

5. Ueda, D., Yamamoto, A., Nishimori, M., et al.: Deep learning for MR angiography: automated detection of cerebral aneurysms. Radiology **290**(1), 187–194 (2018). https://doi.org/10.1148/radiol.2018180901

6. Zhou, Z., et al.: Models genesis: generic autodidactic models for 3D medical image analysis. In: Shen, D., et al. (eds.) MICCAI 2019. LNCS, vol. 11767, pp. 384–393. Springer, Cham (2019). https://doi.org/10.1007/978-3-030-32251-9_42

7. van Griethuysen, J.J.M., et al.: Computational radiomics system to decode the radiographic phenotype. Can. Res. **77**(21), e104–e107 (2017). https://doi.org/10.1158/0008-5472.CAN-17-0339

Cerebral Aneurysm Segmentation

Aν-Net: Automatic Detection and Segmentation of Aneurysm

Suprosanna Shit[1,2(✉)], Ivan Ezhov[1,2], Johannes C. Paetzold[1,2,3], and Bjoern Menze[1,2,4]

[1] Departments of Informatics, Technical University Munich, Munich, Germany
suprosanna.shit@tum.de
[2] TranslaTUM Center for Translational Cancer Research, Munich, Germany
[3] Institute for Tissue Engineering and Regenerative Medicine Helmholtz Zentrum München, Neuherberg, Germany
[4] Department of Quantitative Biomedicine of UZH, Zurich, Switzerland

Abstract. We propose an automatic solution for the CADA 2020 challenge to detect aneurysm from Digital Subtraction Angiography (DSA) images. Our method relies on 3D U-net as the backbone and heavy data augmentation with a carefully chosen loss function. We were able to generalize well using our solution (despite training on a small dataset) that is demonstrated through accurate detection and segmentation on the test data.

Keywords: Aneurysm · Detection · Segmentation

1 Introduction

The leading cause of hemorrhagic stroke is the rupture of intracranial aneurysms. Aneurysms, in general, are vascular anomalies manifested as local *dilation* or balloon-like structure of blood vessels. Identifying intracranial aneurysm in the early stages of its development can reduce the risk of rupture and offer improved treatment planning. However, intracranial aneurysm detection is extremely challenging due to the variability in locations, shapes, and sizes. In clinical practice, different modalities are used for different stages of the diagnostic protocol. For preliminary screening, contrast-agent free modalities, such as TOF MRA or 3DRA, are the crucial and most commonly used modalities. Whereas images using contrast agents, such as DSA images, are used for advanced stages of treatment planning upon requirements.

Earlier approaches to detect cerebral aneurysms rely on 2D image processing, such as sphere enhancing filter [4]. It has been reported [2] that the convolutional neural network (CNN) based aneurysms detection method generalize better and can help radiologists to find more aneurysms without substantially decreasing their specificity. Deep learning-based methods can be classified into two categories, such as 1) global classification and 2) voxel-wise segmentation. While the former [6,16] is more memory and computation efficient, the latter is more useful

© Springer Nature Switzerland AG 2021
A. Hennemuth et al. (Eds.): CADA 2020, LNCS 12643, pp. 51–57, 2021.
https://doi.org/10.1007/978-3-030-72862-5_5

| | Sample 1 | Ground Truth | Sample 2 | Ground Truth |

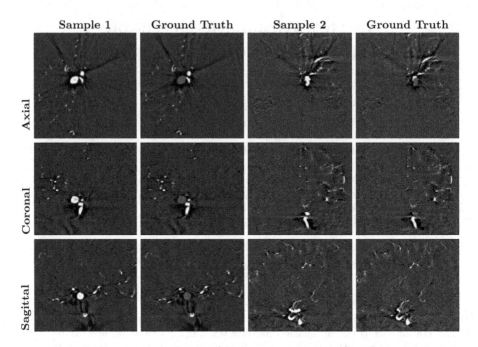

Fig. 1. Few training samples: The first, second, and third row shows the axial, sagittal, and coronal slice of the training examples, respectively. The odd columns show images, and the even columns show the corresponding ground truth annotation. Note the variability in the shape of the aneurysm to be detected by the network.

in providing the exact location and volume of the aneurysm. Several 2D U-net based [13] and 3D DeepMedic based [12] approaches were made to segment the aneurysms. The 3D network gives an extra performance gain at the cost of an increase in computational budget.

Detection and segmentation of cerebral aneurysm at an early stage is critical for clinical treatment planning. Digital subtraction angiography (DSA) is a commonly used modality to identify cerebrovascular pathology. An automatic algorithm to detect aneurysm from DSA images will accelerate the time requirements of the treatment pipeline. Keeping this in mind, we look for an effective solution to the CADA 2020 challenge. Some samples, along with their respective ground truth annotation, are presented in Fig. 1. We identify that the key features that separate aneurysm from a healthy vessel is the local blob-like structure, which can not be differentiated in 2D slices processing. Hence, we opt for a 3D approach. We also identify that given the amount of training samples, solving it as a pure object detection task would be difficult to learn [7]. Alternatively, the segmentation task is much easier to solve, and detection can be a simple post-processing stage. Given the number of aneurysms in a scan can vary, we rely on the assumption that no two aneurysms are adjacent to each other in voxel space and can be separated as different objects from the binary segmentation.

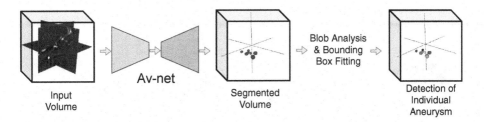

Fig. 2. Schematic overview of our proposed pipeline for detecting aneurysm from DSA images. The $128 \times 128 \times 128$ patch is processed with the $A\nu - net$ to produce the segmentation result. The segmentation result is processed to get rid of boundary artifacts. Subsequently, we do blob analysis and fit bounding boxes around each instance of segmented aneurysms to produce the individual aneurysm label.

Further, to enforce the network to learn shape-based discriminative features, we employ heavy data augmentation as described in Sec. 2.1. An overview of our proposed solution is presented in Fig. 2.

2   Methods

In this short paper, we describe our submission to the CADA 2020 challenge. We implement our segmentation as a three dimensional (3D) binary segmentation. After successful binary segmentation, we count the individual objects, and the final prediction considers one background class ("0") and "n" foreground classes depending on the number of objects in a 3D image volume.

The network architecture is inspired by the encoding-decoding architectures with skip connections. In other words, we use an architecture, which is very similar to the commonly used 3D U-Net architecture with some modifications [1]. We decide to use U-Net, because this is a very successful architecture for diverse medical imaging tasks [3,5,8,10,11,15]. We refer to our solution as Aν-net, which are homophones of 'Aneu-net' and 'an U-net.' The network architecture is depicted in Fig. 3.

Since Dice loss provides an edge at handling class imbalance [8,10] and cross-entropy loss is beneficial for smooth training convergence [9,14], we use both. The total loss function of our method is as follows:

$$\mathcal{L}_{total} = \mathcal{L}_{Dice} + \mathcal{L}_{CrossEnt} \tag{1}$$

2.1   Implementation Details

Our encoder has four-stages with each stage having two residual blocks. We used instance normalization and parametric ReLU as the activation function. As a loss for the network, we used an equally weighted sum of the Dice Loss and the weighted cross entropy loss. The weights for the cross-entropy are [0.01 0.99] for the background and foreground, respectively. We did not use any additional

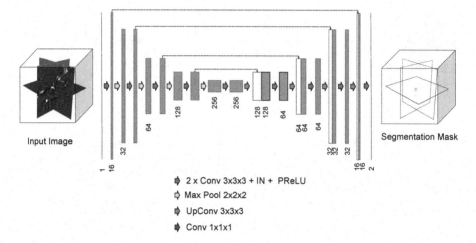

Fig. 3. Schematic overview of our proposed method: Our proposed $A\nu$-*net* architecture is described in detail. The output of the final layer goes through a softmax layer before computing the loss function. For the dice loss, we exclude the background channel and only consider the foreground channel.

training data for this task. We used on the fly data augmentations such as random flipping in all three axes, random [90, 180, 270]-degree rotation along all three axes. We normalized the intensity value of each 3D scan to be of zero mean and unit standard deviation. All networks are implemented in Pytorch using the MONAI package. We use the Adam optimizer with a learning rate of 10^{-3}. The network was trained for 1000 epochs with a batch size of 2 and cubes of 128 voxels per dimension, keeping the configuration of the weight that performs best on the validation set, which split from the training set. The patches were sampled with a 4:1 ratio of the center of the patch being a foreground or background. These strategies help to alleviate the high-class imbalance in the data.

2.2 Detection from Segmentation

We post-process the predicted segmentation of the network to detect and fit the bounding box. We remove any segmentation which is smaller than 150 voxels. We select this threshold from the lower bound of the distribution of aneurysm size of the training data. We also remove artifacts that may appear in the boundary of the image by a simple masking. Subsequently, we do a blob analysis to identify the number of disconnected objects in the prediction. We thereby fit a rectangular bounding box to each instance of the segmented aneurysms.

3   Experiments and Results

We train on 100 scans and validate on 9 scans. We chose our model based on the best dice score on the validation set. We observe a dice score of 82.3 for our best

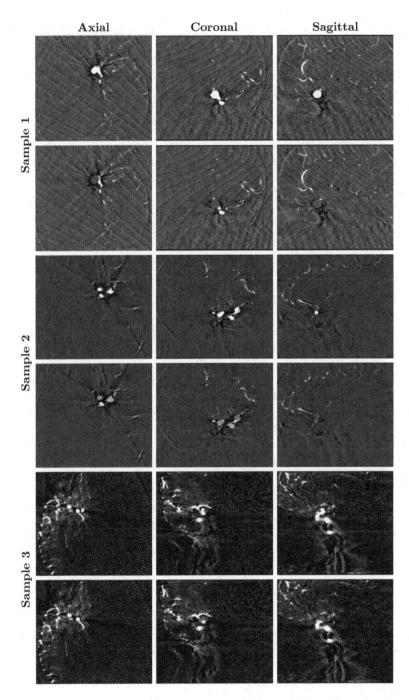

Fig. 4. Qualitative results: The first, second, and third columns represent the axial, sagittal, and coronal slice of the typical sample, respectively, from the test set. The odd rows denote the DSA images, and the even rows show the corresponding prediction from our method. We observe that our model effectively learns to detect multiple aneurysms at the same time.

model on the validation set. However, since the dice score largely varies based on the size of the aneurysm, we suspect that it may have produced a little lower dice score for the test set where most of the aneurysms were medium or small. We attribute this to the fact that during training we did not prioritize small aneurysms over the big ones. Figure 4 shows some qualitative results on the test dataset. Table 1 summarises our performance on the test dataset obtained from the official leaderboard.

Table 1. Our score on the test dataset under the team name IBBM as per the two leaderboards: https://cada.grand-challenge.org/evaluation/leaderboard/ and https://cada-as.grand-challenge.org/evaluation/leaderboard/.

Task	Detection	Segmentation
Score	0.8562	0.6817

4   Conclusions

We provide an effective solution for the CADA 2020 challenge using a simple U-net, without any additional training data and ensemble approach. We achieve accurate segmentation and detection results on all the test cases except a single case where our model does not detect any aneurysm. A minor drawback of our method is that it may struggle to differentiate multiple aneurysms located in one/two voxels' proximity to each other. Nonetheless, our proposed method can serve as a benchmark for developing more complex models aiming to better learn the discriminative anatomical features for aneurysm detection. Specifically, attention module can be used to improve performance on small aneurysms and distinguish aneurysms, which are close apart.

Acknowledgement. Suprosanna Shit and Ivan Ezhov are supported by the Translational Brain Imaging Training Network (TRABIT) under the European Union's 'Horizon 2020' research & innovation program (Grant agreement ID: 765148). Johannes C. Paetzold and Suprosanna Shit are supported by the Graduate School of Bioengineering, Technical University of Munich.

References

1. Çiçek, Ö., Abdulkadir, A., Lienkamp, S.S., Brox, T., Ronneberger, O.: 3D U-Net: learning dense volumetric segmentation from sparse annotation. In: Ourselin, S., Joskowicz, L., Sabuncu, M.R., Unal, G., Wells, W. (eds.) MICCAI 2016. LNCS, vol. 9901, pp. 424–432. Springer, Cham (2016). https://doi.org/10.1007/978-3-319-46723-8_49
2. Duan, H., Huang, Y., Liu, L., Dai, H., Chen, L., Zhou, L.: Automatic detection on intracranial aneurysm from digital subtraction angiography with cascade convolutional neural networks. Biomed. Eng. Online **18**(1), 110 (2019)

3. Gerl, S., et al.: A distance-based loss for smooth and continuous skin layer segmentation in optoacoustic images. In: Martel, A.L., et al. (eds.) MICCAI 2020. LNCS, vol. 12266, pp. 309–319. Springer, Cham (2020). https://doi.org/10.1007/978-3-030-59725-2_30

4. Hentschke, C.M., Beuing, O., Paukisch, H., Scherlach, C., Skalej, M., Tönnies, K.D.: A system to detect cerebral aneurysms in multimodality angiographic data sets. Med. Phys. **41**(9), 091904 (2014)

5. Li, H., et al.: DiamondGAN: unified multi-modal generative adversarial networks for MRI sequences synthesis. In: Shen, D., et al. (eds.) MICCAI 2019. LNCS, vol. 11767, pp. 795–803. Springer, Cham (2019). https://doi.org/10.1007/978-3-030-32251-9_87

6. Nakao, T., et al.: Deep neural network-based computer-assisted detection of cerebral aneurysms in MR angiography. J. Magn. Reson. Imaging **47**(4), 948–953 (2018)

7. Navarro, F., Sekuboyina, A., Waldmannstetter, D., Peeken, J.C., Combs, S.E., Menze, B.H.: Deep reinforcement learning for organ localization in CT. arXiv preprint arXiv:2005.04974 (2020)

8. Navarro, F., et al.: Shape-aware complementary-task learning for multi-organ segmentation. In: Suk, H.-I., Liu, M., Yan, P., Lian, C. (eds.) MLMI 2019. LNCS, vol. 11861, pp. 620–627. Springer, Cham (2019). https://doi.org/10.1007/978-3-030-32692-0_71

9. Paetzold, J.C., et al.: Transfer learning from synthetic data reduces need for labels to segment brain vasculature and neural pathways in 3D. In: International Conference on Medical Imaging with Deep Learning-Extended Abstract Track (2019)

10. Qasim, A.B., et al.: Red-GAN: attacking class imbalance via conditioned generation. Yet another medical imaging perspective. In: Medical Imaging with Deep Learning. PMLR (2020)

11. Shit, S., et al.: clDice–a topology-preserving loss function for tubular structure segmentation. arXiv preprint arXiv:2003.07311 (2020)

12. Sichtermann, T., Faron, A., Sijben, R., Teichert, N., Freiherr, J., Wiesmann, M.: Deep learning-based detection of intracranial aneurysms in 3D TOF-MRA. Am. J. Neuroradiol. **40**(1), 25–32 (2019)

13. Stember, J.N., et al.: Convolutional neural networks for the detection and measurement of cerebral aneurysms on magnetic resonance angiography. J. Digit. Imaging **32**(5), 808–815 (2019)

14. Tetteh, G., Efremov, V., Forkert, N.D., Schneider, M., Kirschke, J., et al.: Deepvesselnet: vessel segmentation, centerline prediction, and bifurcation detection in 3-D angiographic volumes. arXiv preprint arXiv:1803.09340 (2018)

15. Todorov, M.I., et al.: Machine learning analysis of whole mouse brain vasculature. Nat. Methods **17**(4), 442–449 (2020)

16. Ueda, D., et al.: Deep learning for MR angiography: automated detection of cerebral aneurysms. Radiology **290**(1), 187–194 (2019)

3D Attention U-Net with Pretraining: A Solution to CADA-Aneurysm Segmentation Challenge

Ziyu Su[1], Yizhuan Jia[2], Weibin Liao[1], Yi Lv[1], Jiaqi Dou[3], Zhongwei Sun[2(✉)], and Xuesong Li[1(✉)]

[1] School of Computer Science and Technology, Beijing Institute of Technology, Beijing, China
lixuesong@bit.edu.cn
[2] Mediclouds Medical Technology, Beijing, China
szw@mediclouds.cn
[3] Center for Biomedical Imaging Research, Department of Biomedical Engineering, School of Medicine, Tsinghua University, Beijing, China

Abstract. Early detection and accurate segmentation of cerebral aneurysm is important for clinical diagnosis and prevention of rupture, which would be life threatening. 3D images can provide abundant information for characterizing the aneurysm. But the traditional manual segmentation of aneurysms takes lots of time and effort. Therefore, accurate and rapid automatic algorithm for 3D segmentation of aneurysm is needed. U-Net is a widely used deep learning network in medical image segmentation, but its performance is limited by the amount of data. In this challenge of aneurysm segmentation, we proposed to add attention gate and Models Genesis pretraining mechanisms to the classical U-Net model to improve the results. The dice of 3D U-net, 3D Attention U-Net, pretrained 3D U-Net and pretrained 3D Attention U-Net are $0.881, 0.884, 0.890$ and 0.907, respectively. The experimental results show that the use of attention gate and Models Genesis can significantly improve the performance of U-Net model in segmenting aneurysms. This work achieved rank one in CADA 2020- Aneurysm Segmentation Challenge.

Keywords: Image segmentation · 3D Attention U-Net · Transfer learning

1 Introduction

Cerebral aneurysms are local dilations of arterial blood vessel at the weak site of the vessel wall. Subarachnoid hemorrhage (SAH) caused by rupture of a cerebral aneurysm is a life-threatening condition associated with high mortality and morbidity [1]. Automatic three-dimensional aneurysm segmentation can not only greatly reduce the time for reviewing the image, but also has the potential to provide quantitative characterization of the aneurysm and improve the diagnosis accuracy.

Challenge for aneurysm segmentation is the task 2 in CADA 2020. The training set released by the organizers of the CADA Challenge included 109 data sets and 127 annotated aneurysms. Through this competition, the organizers hope that the contestants can find the aneurysm quickly and decide the appropriate rupture prevention strategy

© Springer Nature Switzerland AG 2021
A. Hennemuth et al. (Eds.): CADA 2020, LNCS 12643, pp. 58–67, 2021.
https://doi.org/10.1007/978-3-030-72862-5_6

according to the shape of the aneurysm. This task's final score metrics include running time, Jaccard, Hausdorff distance, average distance, Pearson correlation coefficient, bias and standard deviation of the difference between predicted and reference volumes.

Recently, deep learning methods with good performance have been proposed for the segmentation of cerebral aneurysms [2–4].U-Net network, first published at the MICCAI conference in 2015 by Ronneberger et al. [5], is a widely used deep neural network in medical image segmentation due to its good performance [6, 7], and many U-Shaped semantic segmentation networks have been proposed. Alom et al. [8] used cyclic convolution network and cyclic residual convolution network to improve the accuracy of segmentation networks based on R2U-Net, and tested the model on different medical images such as blood vessels, lungs and skin. Isensee et al. [7] proposed a nnU-Net network that allows high flexibility in terms of the input data set. The nnU-Net can automatically adjust the necessary relevant parameters (e.g. preprocessing, the exact patch size, batch size, and inference settings) according to the characteristics of the data set. The network was shown to perform well in 6 well-established segmentation challenges.

Although U-Net has become a standard network of image segmentation, but for medical images, false positive prediction often occurs when the segmented organs or tissues show great differences among different patients. The current solution is cascaded CNN [14], whose basic idea is to divide the segmentation process into two steps: extracting ROI (region of interest) and segmenting the specific ROI. The disadvantage is that all models in the cascade network will repeatedly extract similar low-level features, resulting in the overuse of model parameters, and high computing resources are needed to train multiple models. Oktay et al. [9] proposed the attention gate (AG) mechanism to suppress learning information of the irreverent regions in the input image and thus, highlight features of the regions of interest. It was shown that integrating AG into U-Net network can significantly improve the sensitivity and prediction accuracy of the model.

Different from natural image processing, medical images processing usually was limited by a small amount of data because of the difficulty in labeling. This would lead to network with poor performance if the training was simply performed on the small amount of data. Transfer learning [10], which starts with a model pretrained with a large dataset, is a popular method to address this issue, as a better starting point for the model generally will lead to better results.

In this work, we aim to use U-Net network to segment aneurysms in DSA (digital subtraction angiography) images. The proposed network is called pretrained 3D Attention U-Net. To improve the performance of the model, the following elements were included in the model:

(1) A 3D U-Net network architecture that mainly focuses on 3D anatomical information was used. In addition, the mechanism of AG was incorporated into the U-Net network to enhance learning of the key features of aneurysms.

(2) A self-supervised learning approach named Models Genesis [11] was used to pretrain the model to learn the prior knowledge of spatial structure characteristics of vessels. This pretraining step is helpful for reducing the dependence of the proposed network on the amount of training data.

In order to assess the performance of our model, we performed aneurysm segmentation with three other U-Net-based models, i.e. 3D U-net, the original 3D Attention U-Net and pretrained 3D U-Net.

2 Method

We used 3D U-Net with AG to segment aneurysm. We first used Models Genesis to pretrain the model to learn the spatial structure characteristics of vessels, which used the transformed images as input and the original images as output. Then we started with the mode obtained from the pretraining and performed transfer learning to further improve our model for the purpose of segmenting aneurysm. In this training step, the original images and the corresponding augmented images were used as input and the labeled images as output.

2.1 Data Preparation

Before the experiment, data augmentation was performed by flipping, rotating, shifting and gray-value rescaling the original images (Table 1).

Table 1. Different data augmentation techniques and the corresponding parameters.

S. no	Data augmentation technique	Parameters
1	Flipping	Top
		Bottom
		Right
		Left
2	Rotating	0°
		−90°
		90°
		180°
3	Shifting	Top
		Bottom
		Right
		Left
4	Gray-value rescaling	0.9
		1.2

Data preparation includes spatial normalization, intensity normalization and cube extraction.

Spatial Normalization. Since the image resolution was not exactly the same across the different datasets, in this step the images were resized to achieve a voxel size of $0.5 \times 0.5 \times 0.5$ mm^3 for all datasets.

Intensity Normalization. The original datasets have large intensity range and the range was even very different between datasets. To mitigate its influence on the performance of the segmentation method, the intensity larger than 2500 was truncated to 2500, and then all intensity values were divided by 2500.

Cube Extraction. To reduce the computational cost, a cube with size of 64 pixels centered at the seed points was extracted for each aneurysm.

2.2 Pretraining

We used a self-supervised learning approach called Models Genesis [11] for pretraining. The principle of the model is to do some image transformation on the original image, and then let the model restore the original image. In this way, the original image is used as the label to supervise the model training, and the model will learn the features directly from the data. The structure of the model is shown in the Fig. 1.

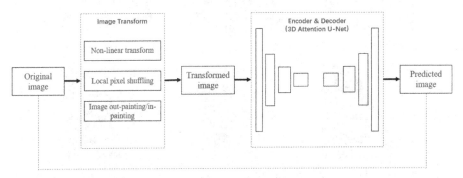

Fig. 1. The structure of Models Genesis. The transformed image is obtained after non-linear transform and local pixel shuffling and image out-painting/in-painting. The images were inputted into the Encoder-Decoder, i.e. 3D Attention U-Net, to restore and learn the spatial structure and intensity information of vascular tissue.

The type of transformation is critical, since it will determine what types of feature the model will learn. For example, the model will learn the appearance features of the image, if a non-linear transform is used. The model will retrieve the global information of the image, including the texture features and edge information of the targeted organs, when local pixel shuffling is used for image transformation. The overall geometry and spatial layout of organs can be learned by the model if image out-painting/in-painting is used for transformation. In this work, all of the three aforementioned transformation methods were used.

The pretraining was performed on a series of small image volumes that were extracted from the original 3D image volumes. Note that to ensure the learning ability of the network, the above data augmentation method was used to increase the number of datasets. In consideration of pulmonary CT images has much more complex feature than cerebral angiography image and similar vessel structure, we pretrained and tested model on several CT images of pulmonary vessels, which were obtained from LNDb: A Lung Nodule Database on Computed Tomography [13]. Example images are shown in Fig. 2.

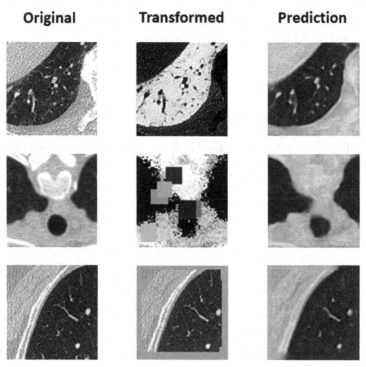

Fig. 2. The first column is the original image, the second column is the transformed image, and the third column is the image restored by the model. Each row corresponds to a type of transformation. In the first row, the nonlinear transformation was used; in the second row, the local pixel reorganization was used; in the third row, the image in-painting was used.

2.3 3D Attention U-Net

We introduced AGs into our U-Net model, to highlight the salient features of skipping connections. The input of the AGs is the feature map transmitted by skipping connection, first zooming using the attention coefficient, and then using the gating vector for each pixel to determine the target area. The attention coefficients identify the region to be segmented and prune the feature responses that is independent of segmentation task. The gating vector contains context information and prune the lower-level feature responses.

During training, AGs can focus on the target structure without additional supervision, and during testing, AGs will immediately generate soft region proposals. The introduced AGs improve the sensitivity and accuracy of the segmentation model by suppressing feature activation in unrelated regions. In this way, the necessity of using external localization model can be eliminated while maintaining high prediction accuracy.

The structure of the model is shown in the Fig. 3.

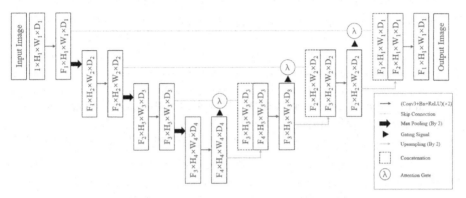

Fig. 3. The structure of the Attention U-Net. Like the original Attention U-Net, our model progressively filters and downsamples the input image at a ratio of 2 times on each scale in the coding part.

In this study, the initial filter number we used was 44. Filter number was doubled after every max pooling layer and halved after every upsampling layer. The convolution kernel size is $3 \times 3 \times 3$. The batch size was 4. We usually trained the models for 100 epochs, and if the validation loss didn't decrease for 15 epochs, training would be stopped.

Adam optimizer was used to update the weights of the network with an initial learning rate of $\alpha = 1 \times 10^{-4}$. Six-fold cross validation was employed, in order to get a reliable and stable model.

For purpose of comparison to our model, we used the same data preparation, loss function and optimizer. And we carried out the same pretraining (if needed), training process. In addition to using the original images as input for training, we also segmented the blood vessels [12] in the image and used the segmentation results as the input to obtain another 3D U-Net model.

2.4 Loss

We used the sum of dice loss and binary cross entropy (BCE) as the loss function during the model training. The reason of using of BCE as a part of loss is that the gradient declines very slow decline in the first stage of training when only using dice loss, and adding BCE can greatly accelerate the gradient decline in this stage.

The dice loss is calculated as Eq. 1, where S refers to the output of our model, and R refers to ground truth and |·| denotes the volume of the region.

$$dice\_loss = 1 - \frac{2|S \cap R|}{|S| + |R|} \tag{1}$$

The BCE is calculated as Eq. 2, where y refers to the output of our model, and $\hat{y}$ refers to ground truth.

$$bce = -(y \cdot log\hat{y} + (1 - y) \cdot log(1 - \hat{y})) \tag{2}$$

3   Experiments and Results

The training was performed on a NVIDIA RTX 2080 Ti GPU with 11 Gb memory. The CPU is AMD Ryzen 5 2600X.

Based on these data and the environment, we performed the following experiments.

3.1   Experiments

We randomly selected 100 images and the corresponding augmented images as the input of the model for training and verification, and the remaining 9 images as the test set.

To better illustrate our method, we conducted ablation experiments to show how some tricks affect the model final performance. We evaluate the model performance under different experimental settings: (1) the original 3D U-Net. (2) 3D Attention U-Net, without pretraining. (3) the original 3D U-Net with the pretraining model (4) 3D Attention U-Net with the pretraining model. In addition, we compared our model with a 3D U-Net which takes segmented vascular trees as input images.

3.2   Result

The indicators of all results on the tested data sets are shown in Table 2. The results show that the separate addition of attention gates and pretraining can improve the accuracy of segmentation, and our model introduce both AG and pretraining to get better results in aneurysm segmentation. For the same U-Net model, the dice obtained by using the original image as input is higher than using segmented blood vessels as input.

Three cases of the segmentation results are shown in Fig. 4.

Table 2. Results of experiments. Both Hausdorff distance (HD) and Average distance (AVD) are in pixels.

Model	Input image	Dice	HD	AVD
3D U-Net	Original image	0.881	4.116	**0.103**
3D U-Net	Segmented vessel images	0.841	13.828	0.968

(continued)

Table 2. (*continued*)

Model	Input image	Dice	HD	AVD
3D attention U-Net	Original image	0.884	8.828	0.541
3D U-Net + pretraining	Original image	0.890	5.375	0.186
3D Attention U-Net + pretraining (ours)	Original image	**0.907**	**3.869**	0.139

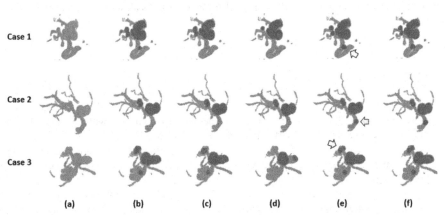

Fig. 4. Three cases of the segmentation results from different models. The pictures above are: (a) vessels from the original image (obtained by using threshold); (b) 3D U-Net; (c) pretrained 3D U-Net; (d) 3D attention U-Net; (e) pretrained 3D attention U-Net (ours); (f) ground truth. Light red represents the background vessels and dark red represents the result predicted by the model. The result of U-Net which uses segmented blood vessels as input has a lot of noises, and the image quality is poor, so it is not shown here. The yellow arrows indicate the advantages of our model. (Color figure online)

4 Discussion and Conclusion

In this paper, we proposed a pre-trained 3D Attention U-Net to segment aneurysms. The attention gate mechanism and Models Genesis pretraining were included in order to reduce the impact of small amount of data on model training. Experiments shows that our model is better than the original 3D U-Net [14], 3D Attention U-Net without pretraining [9] and the original 3D U-Net with the pretraining [11]. Using Models Genesis for pretraining allows us to achieve better results in a limited data set, allowing the model to learn the morphological information of blood vessels in advance. The attention mechanism allows the model to focus more on the aneurysm area during training and reduces the sensitivity of the model to tumor size and shape and improves the prediction accuracy. The combination of the two gives our model the highest score. This suggests that our model is beneficial for addressing the issue of small amount of training data and may be useful in clinical practice.

On the other hand, we only use the segmented blood vessels in the image as the input of the model to optimize the model, but we do not get the expected results. Here we analyze the reason that our vascular segmentation model is not well done. While segmenting the edge of the blood vessel, it is also segmenting the edge of the aneurysm, and in this step, it is easy to destroy the shape of the original aneurysm and other characteristic information. In addition, the U-Net network not only learns the characteristic information of blood vessels, but also optimizes its learning ability by learning the characteristic information of the background. We are unable to follow up further due to time constraints, but we believe that there is still room for improvement in this approach in the future.

References

1. Anker-Møller, T., Hvas, A.M., Sunde, N., et al.: Proteins of the Lectin Pathway of complement activation at the site of injury in subarachnoid hemorrhage compared with peripheral blood. Brain Behav. **10**(8), e01728(2020). https://doi.org/10.1002/brb3.1728
2. Duan, Z., Montes, D., Huang, Y., et al.: Deep Learning Based Detection and Localization of Cerebal Aneurysms in Computed Tomography Angiography. arXiv preprint arXiv:2005.11098 (2020)
3. Jin, H., Geng, J., Yin, Y., et al.: Fully automated intracranial aneurysm detection and segmentation from digital subtraction angiography series using an end-to-end spatiotemporal deep neural network. J. NeuroInterv. Surgery **12**, 1023–1027 (2020). https://doi.org/10.1136/neurintsurg-2020-015824
4. Mohammadi, S., Mohammadi, M., Dehlaghi, V., Ahmadi, A.: Automatic segmentation, detection, and diagnosis of abdominal aortic aneurysm (AAA) using convolutional neural networks and hough circles algorithm. Cardiovasc. Eng. Technol. **10**(3), 490–499 (2019). https://doi.org/10.1007/s13239-019-00421-6
5. Ronneberger, O., Fischer, P., Brox, T.: U-Net: convolutional networks for biomedical image segmentation. In: Navab, N., Hornegger, J., Wells, W.M., Frangi, A.F. (eds.) MICCAI 2015. LNCS, vol. 9351, pp. 234–241. Springer, Cham (2015). https://doi.org/10.1007/978-3-319-24574-4_28
6. Ibtehaz, N., Rahman, M.S.: MultiResUNet: rethinking the U-Net architecture for multimodal biomedical image segmentation. Neural Netw. **121**, 74–87 (2020). https://doi.org/10.1016/j.neunet.2019.08.025
7. Isensee, F., Jaeger, P.F., Kohl, S.A.A., et al.: nnU-Net: a self-configuring method for deep learning-based biomedical image segmentation. Nature Methods, 1–9 (2020). https://doi.org/10.1038/s41592-020-01008-z
8. Zhou, T., Ruan, S., Canu, S.: A review: deep learning for medical image segmentation using multi-modality fusion. Array **3**, 100004 (2019). https://doi.org/10.1016/j.array.2019.100004
9. Oktay, O., Schlemper, J., Folgoc, L.L., et al.: Attention U-Net: learning where to look for the pancreas. arXiv preprint arXiv:1804.03999 (2018)
10. Morid, M.A., Borjali, A., Del Fiol, G.: A scoping review of transfer learning research on medical image analysis using ImageNet. arXiv preprint arXiv:2004.13175 (2020)
11. Zhou, Z., et al.: Models genesis: generic autodidactic models for 3D medical image analysis. In: Shen, D., et al. (eds.) MICCAI 2019. LNCS, vol. 11767, pp. 384–393. Springer, Cham (2019). https://doi.org/10.1007/978-3-030-32251-9_42

12. Zhou, S., et al.: Statistical intensity- and shape-modeling to automate cerebrovascular segmentation from TOF-MRA data. In: Shen, D., et al. (eds.) MICCAI 2019. LNCS, vol. 11765, pp. 164–172. Springer, Cham (2019). https://doi.org/10.1007/978-3-030-32245-8_19
13. Pedrosa, J., et al.: LNDb: a lung nodule database on computed tomography. arXiv:1911.08434
14. Li, H., Lin, Z., Shen, X., Brandt, J., Hua, G.: A convolutional neural network cascade for face detection. Comput. Vis. Pattern Recogn. IEEE (2015). https://doi.org/10.1109/CVPR.2015.7299170

Exploring Large Context for Cerebral Aneurysm Segmentation

Jun Ma[1]([⊠])[ID] and Ziwei Nie[2]

[1] Department of Mathematics, Nanjing University of Science and Technology,
Nanjing, China
`junma@njust.edu.cn`
[2] Department of Mathematics, Nanjing University, Nanjing, China

Abstract. Automated segmentation of aneurysms from 3D CT is important for the diagnosis, monitoring, and treatment planning of the cerebral aneurysm disease. This short paper briefly presents the main technique details of the aneurysm segmentation method in MICCAI 2020 CADA challenge. The main contribution is that we configure the 3D U-Net with a large patch size, which can obtain the large context. Our method ranked second on the MICCAI 2020 CADA testing dataset with an average Jaccard of 0.7593. Our code and trained models are publicly available at https://github.com/JunMa11/CADA2020.

Keywords: Segmentation · Deep learning · U-Net · CT

1 Introduction

Accurate segmentation of aneurysms plays an important role in the quantitative analysis for risk assessment and monitoring of the cerebral aneurysm disease. Manual segmentation is time-consuming and suffers from inter- and intra-observer variability. Thus, fully automatic aneurysm segmentation methods are highly demanded. In MICCAI 2020, Cerebral Aneurysm Segmentation Challenge was hold to benchmark different segmentation methods. The main goal of this short paper is to present the main technique details of our methods. Thus, the medical background of aneurysms and the overview of the state-of-the-art aneurysm segmentation methods are out of the scope of this paper. More details of the challenge background and motivation are available in the challenge design document (http://doi.org/10.5281/zenodo.3715011).

Automatic segmentation of aneurysms is a challenging problem. Figure 1 presents an example of the aneurysm CT image and the corresponding ground truth. It can be found that

- the aneurysm is very small and occupies only a very small part of the whole image;
- other anatomical structures show a similar appearance as the aneurysm, which makes it difficult to distinguish them.

A. Hennemuth et al. (Eds.): CADA 2020, LNCS 12643, pp. 68–72, 2021.
https://doi.org/10.1007/978-3-030-72862-5_7

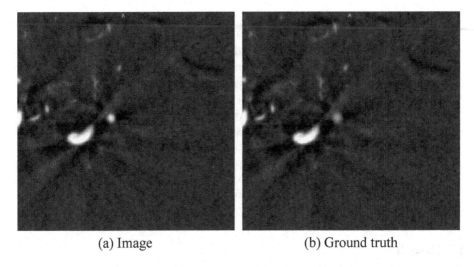

(a) Image   (b) Ground truth

Fig. 1. An example of the aneurysm CT image and the corresponding ground truth.

Our main motivation was from the first-place solution [4] in the MICCAI 2018 brain tumor segmentation challenge (BraTS) where using a larger image patch (i.e., $160 \times 192 \times 128$) as the network input can obtain better performance than a smaller image patch size (even with batch normalization). The potential reason might be that large patch size can provide more semantic context information for the network.

2   Dataset and Method

The challenge organizers provided 110 cases for training and 22 cases for testing. Specifically, image data of patients with cerebral aneurysms without vasospasm were collected for diagnostic and treatment decision purposes [3]. The image data were acquired utilizing the digital subtraction AXIOM Artis C-arm system using a rotational acquisition time of 5s with 126 frames. Post-processing was performed using LEONARDO InSpace 3D (Siemens, Forchheim, Germany). A contrast agent (Imeron 300, Bracco Imaging Deutschland GmbH, Germany) was manually injected into the internal carotid (anterior aneurysms) or vertebral (posterior aneurysms) artery. Reconstruction of a volume of interest selected by a neurosurgeon generated a stack of 220 image slices with matrices of 256×256 voxels in-plane, resulting in an isotropic voxel spacing of $0.5 \times 0.5 \times 0.5 \, \mathrm{mm}^3$.

In our solution, 3D nnU-Net [1,2] was used as the main network architecture. In particular, we modified the default nnU-Net with a large image patch size of $192 \times 224 \times 192$ and a batch size of 2 as the input, which were the largest patch size and batch size that the GPU memory allowed. The detailed settings are as follows:

- Preprocessing. We use three-order spline interpolation to resample all the images to a common spacing of $0.5429 \times 0.5429 \times 0.5429 \, \mathrm{mm}^3$, and normalize the intensity to a mean of 0 and standard deviation of 1.
- Training. The U-Net has six resolutions. The feature size is decreased by half in each resolution via downsampling with strided convolutions. The optimizer is stochastic gradient descent with an initial learning rate (0.01) and a Nesterov momentum (0.99). To avoid overfitting, standard data augmentation techniques are used during training, such as rotation, scaling, adding Gaussian noise, and gamma correction. The loss function is the unweighted sum of Dice loss and cross entropy. We apply five-fold cross validation with 110 training cases. Each fold is trained on a NVIDIA TITAN V100 GPU.
- Inference. The 5 models resulting from training are used as an ensemble for predicting the test cases.

3   Results

Table 1 shows the quantitative results of the five-fold cross-validation. The performances vary among different folds, indicating that the training cases have different difficulties. Our method achieves an average Jaccard of 0.8112, Dice of 0.8861, precision of 0.8934 and recall of 0.9036 in cross validation. Figure 2 presents some visualized segmentation results. Overall, the segmentation results are accurate. However, small segmentation errors can significantly degenerate the Jaccard/Dice scores because the aneurysms are very small.

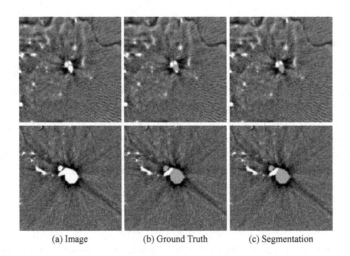

(a) Image   (b) Ground Truth   (c) Segmentation

Fig. 2. Visualized examples in the cross-validation segmentation results.

We apply the proposed method on the hidden testing set where the ground truth are held by the organizers. The total inference time is 88 min. Table 2

Table 1. Quantitative results of five-fold cross-validation results on the training set. AVG denotes the average results of all training cases.

Fold	Jaccard	Dice	Precision	Recall
0	0.7901	0.8737	0.8970	0.8742
1	0.8335	0.9034	0.9046	0.9173
2	0.7966	0.8805	0.8611	0.9163
3	0.7718	0.8470	0.8661	0.8904
4	0.8638	0.9256	0.9384	0.9197
AVG	0.8112	0.8861	0.8934	0.9036

Table 2. Quantitative results on the testing set.

Metric	Value
Jaccard	0.759
Volume bias (mm^3)	75.8
Mean distance (mm)	3.54
Volume pearson R	0.998
Hausdorff distance (mm)	4.97

present the quantitative segmentation results. It can be found that the average Jaccard score is 0.7593 which is lower than the cross validation results. We noticed that the minimal Jaccard score is 0, indicating that there are some very challenging cases that our method fails to handle.

4 Conclusion

In this short paper, we present our segmentation method for cerebral aneurysm segmentation. Specifically, we employed the well-known 3D U-Net and configured with a large patch size. This simple method achieved the second place with a Jaccard score of 0.7593 on the testing set (https://cada-as.grand-challenge.org/FinalRanking/). In future, we will speed up the inference time to make our method more efficient.

Acknowledgement. This work is supported by the National Natural Science Foundation of China (No. 11531005, No. 11971229). We are grateful to the High Performance Computing Center of Nanjing University for supporting the blade cluster system to run the experiments. We also highly appreciate the CADA organizers for holding the great challenge and creating the publicly available dataset.

References

1. Çiçek, Ö., Abdulkadir, A., Lienkamp, S.S., Brox, T., Ronneberger, O.: 3D u-net: learning dense volumetric segmentation from sparse annotation. In: International Conference on Medical Image Computing and Computer-assisted Intervention, pp. 424–432 (2016)
2. Isensee, F., Jäeger, P.F., Kohl, S.A.A., Petersen, J., Maier-Hein, K.H.: NNU-net: a self-configuring method for deep learning-based biomedical image segmentation. Nat. Methods **18**(2), 203–211 (2021)
3. Ivantsits, M., et al.: Cerebral aneurysm detection and analysis challenge 2020 (CADA) (2021)
4. Myronenko, A.: 3D MRI brain tumor segmentation using autoencoder regularization. In: International MICCAI Brainlesion Workshop, pp. 311–320 (2018)

Cerebral Aneurysm Rupture Risk Estimation

CADA Challenge: Rupture Risk Assessment Using Computational Fluid Dynamics

Kartik Jain$^{(\boxtimes)}$ (ID)

Faculty of Engineering Technology, University of Twente,
P.O. Box 217, 7500 Enschede, AE, The Netherlands
k.jain@utwente.nl
https://jainka.github.io

Abstract. The phase 3 of the cerebral aneurysm detection and analysis (CADA) challenge involved rupture risk estimation of intracranial aneurysms using computational methods. In this work we performed computational fluid dynamics (CFD) on a subset of aneurysm cases provided by the challenge committee. A large number of aneurysm cases were available, CFD analysis using the lattice Boltzmann method (LBM) were performed on 18 of them. These 18 aneurysms were chosen on the basis of most distinct shape, size and location. Direct numerical simulations were performed to identify wall shear stress and pressure, and associate these hemodynamic quantities with the rupture status of aneurysms and eventually extrapolate those findings to other aneurysms. The results of the DNS may serve as inputs for data driven methods to identify qualitative maps of hemodynamic quantities in aneurysms. In this article we report the results of CFD and discuss hypotheses associating the flow characteristics with the rupture risk of aneurysms.

Keywords: Intracranial aneurysm · Computational fluid dynamics · Lattice boltzmann method

1 Introduction

The cerebral aneurysm detection and analysis (CADA)[1] challenge was an international scientific initiative that was launched to further our insights into the rupture risk estimation of aneurysms using computational methods. The challenge specifically focused on the automation of analyses that are usually done for the estimation of rupture risk in aneurysms. For example, computational fluid dynamics (CFD) has evolved as an important tool for the estimation of hemodynamic forces that act on an aneurysm thereby leading to its rupture [1–3]. The CFD computations estimate the wall shear stress (WSS), pressure and velocity field inside an aneurysm and identify how an interplay of these forces might

---

[1] https://cada.grand-challenge.org.

© Springer Nature Switzerland AG 2021
A. Hennemuth et al. (Eds.): CADA 2020, LNCS 12643, pp. 75–86, 2021.
https://doi.org/10.1007/978-3-030-72862-5_8

lead to the rupture. No general agreement however exists whether it is high or low WSS that leads to rupture of an aneurysm. See for example, Chien et al. [4] advocates that high WSS must lead to the rupture of an aneurysm whereas Shojima et al. [5] suggests that the low WSS is a cause of aneurysm rupture. Hassan et al. [6] on the other hand indicates how high systolic pressure in the dome may trigger growth of an aneurysm.

An accurate assessment of hemodynamic quantities in a complex aneurysm requires that the CFD is conducted with sufficiently high resolutions in space and time to accurately capture the intricate features of the flow [7–9] thus requiring a long calculation time and the deployment of high performance computing resources. Motivated by the aforesaid issues and limitations there is a strong need that the CFD practices are idealized on one hand and the speed of rupture risk identification is improved to allow decision making in clinical practise. A time consuming CFD calculation is acceptable if the research is of purely academic nature to understand the etiology and pathophysiology of aneurysms. Several challenges have been organized in the past years by the aneurysm researchers [10–14] to address this very issue. These challenges try to identify the differences between computations from various groups, in particular their association with the rupture risk of aneurysms. The recent CADA challenge has been the biggest effort in this direction so far, which provided an enormous amount of datasets containing information of about 115 aneurysm patients. A detailed information about the CADA dataset is available in Tabea Kossen et al. [15]. The phase 1 and 2 of the challenge concerned morphological and segmentation aspects of the aneurysms whereas the phase 3 concerned classification of the aneurysms based on rupture risk.

This work has been conducted as part of the phase 3 of the challenge. We carried out direct numerical simulations on 18 aneurysms. These aneurysms were selected on the basis of most distinct shapes, location and morphology. The rupture risk was then associated with wall shear stress and pressure obtained from CFD computations, in addition to the shape of the aneurysms. Since the phase 3 of the challenge emphasized *automation* of the procedures, conducting CFD on all the aneurysms was not only cumbersome but less valuable for the challenge. We thus intended to use the CFD results from these aneurysms to train a convolution neural network that could then estimate WSS and pressure in all the aneurysms thereby serving as a tool for further studies. While the neural network framework did not turn out as planned, the detailed DNS on 18 aneurysms provided fairly good insight into wall shear stress and pressure patterns that could be associated with rupture status. DNS were particularly chosen so that the CFD results are as accurate as possible thereby not amplifying foreseeable errors in the neural network outcomes. Also, the intention of the CFD studies was to extend the database of the CADA so that the simulations here could potentially serve as inputs for further research using the data of this challenge. In this article we discuss the physics identified from the simulations and discuss hypotheses that may associate hemodynamic parameters with rupture risk of aneurysms.

2   Methods

The aneurysm geometries were available in the form of surface meshes in STL format. The complexity of the geometries ranged from low to extremely high with some cases harboring multiple aneurysms while others with the inclusion of larger vasculature. In order to ease the mesh generation of such complex cases and to continue with the intention of fully resolved direct numerical simulations (DNS) to identify complex flow features, we chose the lattice Boltzmann method (LBM) for simulations. The structured grid required for LBM computations and its scalability on high performance supercomputers makes it a suitable method when complex anatomy is involved and detailed flow features are of interest. Simulations were conducted using the end to end parallel simulation framework APES (adaptable poly engineering simulator) [16,17]. Computational meshes were created using the mesh generator *Seeder* [18]. The initial conditions were set to zero velocity and pressure. At the outlet a zero pressure extrapolation boundary condition was prescribed as described in Junk and Yang [19]. All the vascular walls were assumed to be rigid and the arbitrary curved surfaces were represented by a higher order no-slip boundary condition, which aligns the LBM velocity directions to the curved surfaces thereby compensating for the staircase approximation of boundaries in a LBM cartesian mesh [20]. The D3Q19 stencil was chosen in this study for LBM computations which describes 19 discrete velocity links per compute cell in 3 dimensions.

The blood rheology was represented with constant density and viscosity using a Newtonian description, i.e. $\rho = 1.025\,g/cm^3$ and $\nu = 0.00345$ Pa s. The peak systolic mean velocity in the middle cerebral artery M1 segment is between $0.4 - 0.5\,\text{m/s}$ [21]. In the basilar artery the peak velocity ranges from 0.4 to $0.8\,\text{m/s}$ [22]. A uniform parabolic velocity profile with a mean of $\bar{u} = 0.4\,\text{m/s}$ was chosen as boundary condition at the inlet to enable a comparison of flow among all aneurysms, and in particular because exploring any transitional characteristics of flow was not the focus of this study. The simulations were executed for 2 s and the characteristics were analyzed at t = 2s. Based on a previous comprehensive mesh convergence study [7,23], the meshes were discretized with a spatial and temporal resolution of $\Delta x = 32\mu m$ and $\Delta t = 1\mu s$ respectively. This resulted in mesh sizes between 25 million and 230 million cells. The Lattice Boltzmann flow solver *Musubi* [24] was used for simulations. In order to ensure the quality of the DNS, we computed the Kolmogorov microscales using the fluctuating component of the strain rate as previously described in Junk and Yang [7,23]. A comparison of the employed resolutions against the computed Kolmogorov scales ensured the quality of the resolutions chosen for the DNS. Simulations were executed using between 240 and 4800 CPUs of the Cartesius system installed in THE NETHERLANDS. Each simulation required between 40 minutes to 10 hours of execution time depending on the number of lattice cells and CPUs. Further information about CPU hours consumption can be found in Klimach et al.[17]. The Cartesius is a heterogenous system made up of several types of nodes like Haswell, Broadwell, Xeon Phi and Sandy Bridge. For the purpose of this work simulations were conducted on Haswell (Intel Xeon Processor E5 v3 Family) nodes to ensure homogeneity. The Cartesius system is managed by the SURF organization[2] in THE NETHERLANDS.

3   Results

The Kolmogorov length, time and velocity scales were computed at locations inside the aneurysm where there were minor fluctuations in the strain rate. Only in the unruptured aneurysm $a135$ some minor fluctuations were found whereas the flow regime in all other cases was laminar. The Kolmogorov length, time and velocity scales were respectively $\eta = 18.86\,\mu\text{m}$, $\tau_\eta = 7.32\,\mu\text{s}$ and $u_\eta = 1.66\,\text{m/s}$. This implies that the ratio of employed resolutions in the simulations to corresponding length and time scales from the Kolmogorov theory were $l^+ = \frac{\Delta x}{\eta} = 1.69$ and $t^+ = \frac{\Delta t}{\tau_\eta} = 0.13$ – thus advocating the sufficiency of employed resolutions for an accurate DNS of the flow in aneurysms.

Figure 1 shows the profiles of wall shear stress at t = 2 s of the simulation in 18 aneurysms. Corresponding profiles of pressure are shown in Fig. 2. The first page (11 cases) in each figure are the unruptured cases while the last 6 are ruptured ones. The rupture status of the last case ($a095$) is not known. The shear stress is both high and low in saccular areas, see for example $a003$ and $a045$ in which the flow impinges the sac of the aneurysms. In $a074$, $a133$ and $a135$ on the other hand the wall shear stress shows elevated profiles in the aneurysm dome itself. In almost all the unruptured cases the high shear zones are confined to the inlet and outlet branches while the distribution is more even in the aneurysm dome. Cases $a098$ and $a135$ are the only ones in which there is a direct impingement of the jet on the aneurysm dome leading to elevated and uneven wall shear stress profiles in the dome. In the ruptured cases on the other hand (Fig. 1l–1q) the wall shear stress is relatively higher and uneven in almost all the cases. In aneurysms $a074$, $a133$ and $a138$ the flow jet impacts the aneurysm dome directly due to the manifestation of the aneurysm at a bifurcation leading to higher WSS and uneven profiles.

The pressure profiles demonstrate similar patterns whereby there is a decrease in pressure from the inlet to the outlets. The morphology of the inlet branch influences the pressure inside the dome remarkably, see for example $a003$ and $a108$ where the tortuos and stenosed inflows result in non-linear pressure distributions already at the inflow. The pressure in the ruptured cases (Fig. 2l–2q) does not provide any distinctive features that could differentiate the ruptured cases from the unruptured ones.

4   Discussion

The DNS on 18 aneurysms in this work is an attempt to identify if the patterns of pressure and WSS as well as the shape of the aneurysms can provide a distinct *marker* that could provide insights into aneurysm rupture risk. Due

---

[2] www.surfsara.nl.

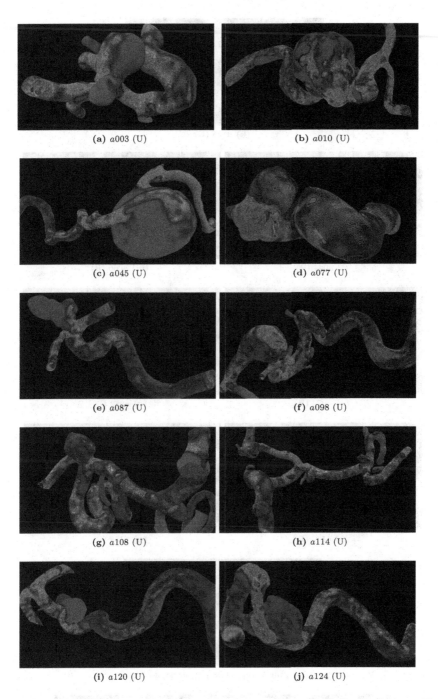

(a) $a003$ (U)

(b) $a010$ (U)

(c) $a045$ (U)

(d) $a077$ (U)

(e) $a087$ (U)

(f) $a098$ (U)

(g) $a108$ (U)

(h) $a114$ (U)

(i) $a120$ (U)

(j) $a124$ (U)

Fig. 1. Wall shear stress magnitudes in 18 aneurysms at $t = 2\,s$ of the simulations. The first page shows the unruptured cases (U) whereas the ruptured cases (R) are shown in the following page. The rupture for case $a095$ is not known.

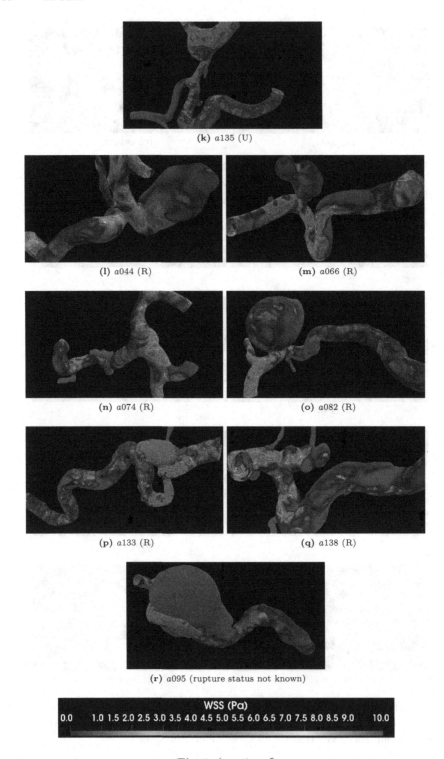

(k) $a135$ (U)

(l) $a044$ (R)

(m) $a066$ (R)

(n) $a074$ (R)

(o) $a082$ (R)

(p) $a133$ (R)

(q) $a138$ (R)

(r) $a095$ (rupture status not known)

WSS (Pa)

0.0   1.0 1.5 2.0 2.5 3.0 3.5 4.0 4.5 5.0 5.5 6.0 6.5 7.0 7.5 8.0 8.5 9.0   10.0

Fig. 1. (*continued*)

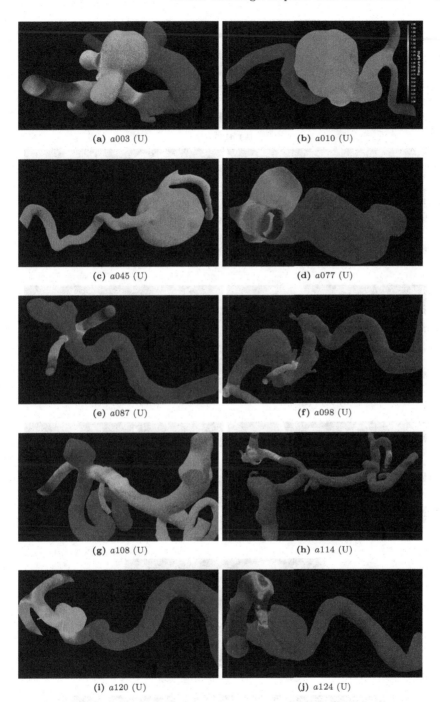

Fig. 2. Pressure profiles in 18 aneurysms at t = 2 s of the simulations. The first page shows the unruptured cases (U) whereas the ruptured cases (R) are shown in the following page. The rupture for case a095 is not known. *Note the different legend bar for cases a010 and a135.*

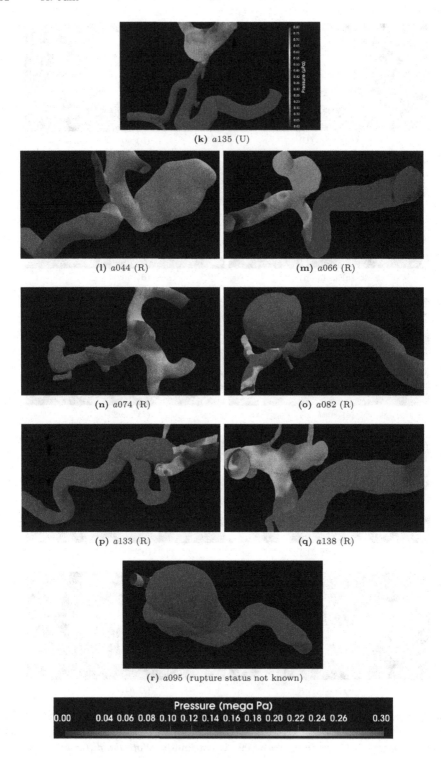

(k) *a*135 (U)

(l) *a*044 (R)

(m) *a*066 (R)

(n) *a*074 (R)

(o) *a*082 (R)

(p) *a*133 (R)

(q) *a*138 (R)

(r) *a*095 (rupture status not known)

Fig. 2. (*continued*)

to the small number of aneurysms that were simulated no distinct feature can be quantitatively identified. In fact, several previous studies that did perform CFD on a large number of cases are also inconclusive [4,5] as aneurysm initiation and rupture is a complex biological process. In fact it has been argued that aneurysms might be a spectrum of diseases [25].

Nonetheless, several hypotheses emanate from the simulations and the availability of the immense data in the database of CADA opens up opportunities for several analyses. Here we first discuss how the simulation results may associate to rupture risk and then associate the findings with a few other aneurysm cases that were provided with the CADA dataset.

Association with Rupture Risk

The mean WSS inside the domes of unruptured cases was between 7.5–10% smaller than the surrounding vasculature in all the unruptured cases except $a098$ and $a135$. In the ruptured cases the mean WSS was higher than the surrounding vasculature by about 4–6%, except for the case $a066$ where it was relatively equal. This observation leads to a hypothesis that the domes of unruptured aneurysms receive relatively less flow that reflects in the lower WSS. A closer look however reveals that the WSS in the surrounding vasculature is much more *uneven* than in the aneurysms themselves. The case $a114$ for example harbors as many as three aneurysms in one patient and relatively large parts of the surrounding vasculature are provided in this geometry. All three aneurysms in this case are unruptured. On the other hand the ruptured cases $a074$, $a133$ and $a138$ do not have much vasculature around in the models and hence simulations are *biased* in a way as a detailed comparison cannot be made. It is the manifestation of these aneurysms at bifurcation that leads to higher flow impingement, high WSS and hence rupture.

Similar observations are clearly drawn from the pressure profiles. One distinct feature that the pressure profiles highlight is the uneven and varying diameter of inlet vessel in many cases that leads to adverse pressure gradients even before the flow reaches the aneurysm sac. It is arguable if these effects are a consequence of the anatomy as not much vasculature is included in the dataset. This question might be answered by the phase 1 of the CADA challenge which aims to quantify the errors introduced by the segmentation of images to obtain the vessel.

Extrapolation of Simulation Results to other Aneurysm Cases

Shape has been suggested as one of the marker for aneurysm rupture [26]. From the present study also it appears that WSS and pressure might be merely *indirect* markers for rupture risk, and the shape of not only the aneurysm but the surrounding vasculature might have a bigger role to play. The aneurysms for CFD were chosen based on a distinctive or representative shape and an attempt on associating rupture risk can be made based on the shape (even independent of the CFD). The shape of $a003$ for example correlates well with $a027$ both of which are unruptured aneurysm. Similarly, the unruptured case $a135$ relates well with $a009$ and $a012$. Similarly the ruptured case, $a138$ has a good shape correlation with $a017$ and $a029$. The ruptured case $a082$ has a similar sidewall structure as $a041$.

The shape of aneurysms can be well quantified by data driven approaches, see for example the radiomics approach recently proposed by Juchler et al. [27]. Future works may use a combination of CFD and shape analysis techniques to arrive at robust and fast analyses of aneurysm rupture risk.

Limitations

This work suffers several limitations. Primarily, the non-inclusion of lumen elasticity is considered as one of the major drawbacks as simulations in rigid aneurysm vessels have already matured at a level when not much new can be investigated. Furthermore, originally 30 total aneurysms were selected from the database for simulations. The 12 that are not reported are the ones that had either more than 5 outflows or they had a small side branch. Clipping of the side branch or prescription of an artificial boundary condition in such aneurysm cases would have increased the amount of uncertainties thus leading to little or no scientific output. Also, the original plan was to perform transient simulations as a second step after these simulations, which was abandoned as the deep learning framework did not work out as planned. Finally, the presence or absence of larger vasculature in some aneurysms poses a *bias* in the simulations as the flow physics gets dramatically altered. Investigation of these aspects in detail is left for future efforts.

References

1. Chung, B., Cebral, J.R.: CFD for evaluation and treatment planning of aneurysms: review of proposed clinical uses and their challenges. Ann. Biomed. Eng. **43**(1), 1–17 (2014)
2. Xiang, J., et al.: Hemodynamic-morphologic discriminants for intracranial aneurysm rupture. Stroke **42**(1), 144–152 (2011)
3. Lu, G., et al.: Influence of hemodynamic factors on rupture of intracranial aneurysms: patient-specific 3D mirror aneurysms model computational fluid dynamics simulation. Am. J. Neuroradiol. **32**(7), 1255–1261 (2011)
4. Chien, A., Tateshima, S., Castro, M., Sayre, J., Cebral, J., Vinuela, F.: Patient-specific flow analysis of brain aneurysms at a single location: comparison of hemodynamic characteristics in small aneurysms. Med. Biol. Eng. Comput. **46**(11), 1113–1120 (2008)
5. Shojima, M., et al.: Magnitude and role of wall shear stress on cerebral aneurysm: computational fluid dynamic study of 20 middle cerebral artery aneurysms. Stroke **35**(11), 2500–2505 (2004)
6. Hassan, T., et al.: Computational simulation of therapeutic parent artery occlusion to treat giant vertebrobasilar aneurysm. Am. J. Neuroradiol. **25**(1), 63–68 (2004)
7. Jain, K., Roller, S., Mardal, K.-A.: Transitional flow in intracranial aneurysms-a space and time refinement study below the Kolmogorov scales using lattice Boltzmann method. Comput. Fluids **127**, 36–46 (2016). https://doi.org/10.1016/j.compfluid.2015.12.011

8. Valen-Sendstad, K., Steinman, D.A.: Mind the gap: impact of computational fluid dynamics solution strategy on prediction of intracranial aneurysm hemodynamics and rupture status indicators. Am. J. Neuroradiol. **35**(3), 536–543 (2013)
9. Dennis, K.D., Kallmes, D.F., Dragomir-Daescu, D.: Further discussion of "cerebral aneurysm blood flow simulations are sensitive to basic solver settings". J. Biomech. **61**, 281–282 (2017)
10. Radaelli, A.G., et al.: Reproducibility of haemodynamical simulations in a subject-specific stented aneurysm model-a report on the virtual intracranial stenting challenge 2007. J. Biomech. **41**(10), 2069–2081 (2008)
11. Steinman, D.A., et al.: Variability of computational fluid dynamics solutions for pressure and flow in a giant aneurysm: the ASME 2012 summer bioengineering conference CFD challenge. J. Biomech. Eng. **135**(2), 021016 (2013)
12. Janiga, G., Berg, P., Sugiyama, S., Kono, K., Steinman, D.A.: The computational fluid dynamics rupture challenge 2013–Phase I: prediction of rupture status in intracranial aneurysms. Am. J. Neuroradiol. **36**(3), 530–536 (2015)
13. Berg, P., et al.: The computational fluid dynamics rupture challenge 2013-phase II: variability of hemodynamic simulations in two intracranial aneurysms. J. Biomech. Eng. **137**(12), 121008 (2015)
14. Valen-Sendstad, K., et al.: Real-world variability in the prediction of intracranial aneurysm wall shear stress: the 2015 international aneurysm CFD challenge. Cardiovasc. Eng. Technol. **9**(4), 544–564 (2018). https://doi.org/10.1007/s13239-018-00374-2
15. Kossen, C.T., et al.: Cerebral aneurysm detection and analysis. (2020). https://doi.org/10.5281/zenodo.3715012
16. Roller, S., et al.: An adaptable simulation framework based on a linearized octree. In: Resch, M., Wang, X., Bez, W., Focht, E., Kobayashi, H., Roller, S. (eds.) High Performance Computing on Vector Systems 2011. Springer, Heidelberg (2011). https://doi.org/10.1007/978-3-642-22244-3_7
17. Klimach, H., Jain, K., Roller, S.: End-to-end parallel simulations with apes. Parallel Comput. Accelerating Comput. Sci. Eng. (CSE) **25**, 703–711 (2014)
18. Harlacher, D.F., Hasert, M., Klimach, H., Zimny, S., Roller, S.: Tree based voxelization of STL data. In: Resch, M., Wang,, X., Bez, W., Focht, E., Kobayashi, H., Roller, S. (eds.) High Performance Computing on Vector Systems 2011. Springer, Heidelberg (2011). https://doi.org/10.1007/978-3-642-22244-3_6
19. Junk, M., Yang, Z.: Asymptotic analysis of lattice Boltzmann outflow treatments. Commun. Comput. Phys. **9**(5), 1117–1127 (2011)
20. Bouzidi, M.H., Firdaouss, M., Lallemand, P.: Momentum transfer of a Boltzmann-lattice fluid with boundaries. Phys. Fluids **13**, 3452 (2001)
21. Krejza, J., et al.: Age and sex variability and normal reference values for the vmca/vica index. Am. J. Neuroradiol. **26**(4), 730–735 (2005)
22. Jain, K., Jiang, J., Strother, C., Mardal, K.-A.: Transitional hemodynamics in intracranial aneurysms - comparative velocity investigations with high resolution lattice Boltzmann simulations, normal resolution ANSYS simulations and MR imaging. Med. Phys. **43**, 6186–6198 (2016). https://doi.org/10.1118/1.4964793. PMID:27806613
23. Jain, K.: Efficacy of the FDA nozzle benchmark and the lattice Boltzmann method for the analysis of biomedical flows in transitional regime. Med. Biol. Eng. Comput. **58**, 1817–1830 (2020). https://doi.org/10.1007/s11517-020-02188-8. PMID:32507933
24. Hasert, M., et al.: Complex fluid simulations with the parallel tree-based lattice Boltzmann solver Musubi. J. Comput. Sci. **5**(5), 784–794 (2014)

25. Strother, C.M., Jiang, J.: Intracranial aneurysms, cancer, X-rays, and computational fluid dynamics. Am. J. Neuroradiol. **33**(6), 991–992 (2012)
26. Raghavan, M.L., Ma, B., Harbaugh, R.E.: Quantified aneurysm shape and rupture risk. J. Neurosurg. **102**(2), 355–362 (2005)
27. Juchler, N., et al.: Radiomics approach to quantify shape irregularity from crowd-based qualitative assessment of intracranial aneurysms. Comput. Method Biomech. Biomed. Eng. Imaging Visual. **8**(5), 538–546 (2020)

Cerebral Aneurysm Rupture Risk Estimation Using XGBoost and Fully Connected Neural Network

Yanfei Liu[1,2], Yunqiao Yang[1,3], Yi Lin[1(✉)], Yuexiang Li[1], Dong Wei[1], Kai Ma[1], and Yefeng Zheng[1]

[1] Tencent Jarvis Lab, Shenzhen, China
ianylin@tencent.com
[2] Hunan University, Changsha, China
[3] Huazhong University of Science and Technology, Wuhan, China

Abstract. Subarachnoid hemorrhage (SAH) caused by the rupture of cerebral aneurysm is a serious life-threatening disease. Therefore, estimating the risk of cerebral aneurysm rupture is clinically important. In this paper, a semi-automatic method for estimating the risk of rupture of cerebral aneurysm was proposed. We applied a variety of methods to extract features of cerebral aneurysm images and 3D modeling, and used XGBoost and fully connected neural network for classification and analysis respectively. The method achieved an F2-score of 0.862 on the test set of CADA 2020.

Keywords: Risk estimation · Classification · Feature extraction

1 Introduction

Cerebral aneurysm is a saccular malformation formed due to the destruction of the intracranial artery wall or the abnormal expansion of the local lumen due to local congenital defects. Once the intracranial aneurysm ruptures, subarachnoid hemorrhage (SAH) will occur [4]. Cerebral aneurysm is the main cause of spontaneous subarachnoid hemorrhage, accounting for 70%–85%. Therefore, the prevention, inspection and treatment of cerebral aneurysms have always been research hot-spots. In recent years, medical imaging methods such as Computed Tomography (CT) and Magnetic Resonance Imaging (MRI) have developed rapidly, which play an important role in assessing the risk of cerebral aneurysm rupture. The consequences of rupture of cerebral aneurysms are very serious. Brain surgery and interventional treatment also have certain risks. A comprehensive assessment of the rupture risk of unruptured aneurysms is of great significance for guiding the treatment of unruptured aneurysms. Evaluating the risk

---

Y. Liu and Y. Yang—The two authors contributed equally to this work.

© Springer Nature Switzerland AG 2021
A. Hennemuth et al. (Eds.): CADA 2020, LNCS 12643, pp. 87–92, 2021.
https://doi.org/10.1007/978-3-030-72862-5_9

factors of the risk of cerebral aneurysm rupture is the key to guiding the clinic in choosing the timing of surgery, preventing aneurysm rupture and bleeding, and reducing mortality and disability.

In the existing literature, many researchers considered that the risk of cerebral aneurysm rupture is related to the aneurysm's morphology, hemodynamics, the level of inflammation in the vessel wall and the patient-specific factors. In terms of morphology, the factors affecting the risk of intracranial aneurysm rupture are the size [2,5], shape [1,6] and location [3] of the aneurysm.

In this paper, we proposed extraction methods based on aneurysm morphological features and image features, and applied XGBoost and fully connected network for classification and analysis.

2   Methods

2.1   Feature Extraction

In order to better estimate the risk of cerebral aneurysm rupture, a total of 110 features are extracted from aneurysm morphology, data distribution and a CNN-based method.

Morphological Features. By manually measuring the three-dimensional modeling files of cerebral aneurysms and vessels, the neck width, the maximum diameter and the length of each cerebral aneurysm are obtained, as shown in Fig. 1(a), and the ratio of the maximum diameter to the length is calculated.

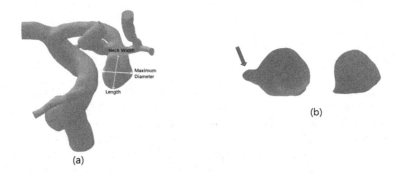

Fig. 1. The measurements made in the STL file. (a) is the 3D modeling of the vessel, in which the neck width, length, maximum diameter of cerebral aneurysm are measured. (b) is the 3D modeling of the cerebral aneurysm. The image on the left has a protrusion at the arrow.

The small globular protrusions and lobulated state are marked, see Fig. 1(b). The cerebral aneurysms which is similar to the left are labeled by 1.

Data Distribution Features. The images of the cerebral vessel are sampled based on the bounding box of the mask of cerebral aneurysm. For the sampled patches, the morphological features include the following: the first order features (the distribution of voxel intensities within the image region defined by the mask through commonly used and basic metrics), the shape features (the three-dimensional size and shape of the ROI), the gray level co-occurrence Matrix (the second-order joint probability function of an image region constrained by the mask), the gray level size zone (the number of connected voxels that share the same gray level intensity), the gray level run length matrix (the length in number of pixels, of consecutive pixels that have the same gray level value), the neighboring gray tone difference matrix (the difference between a gray value and the average gray value of its neighbors within distance), and the gray level dependence matrix (the number of connected voxels within distance that are dependent on the center voxel). These features are extracted by a open-source feature extractor, PyRadiomics (http://pyradiomics.readthedocs.io).

Features Extracted by CNN. A CNN-based binary classification network is applied to pre-classify the entire images and their masks, that is, to reduce the dimension of the images and their masks, as shown in Fig. 1(b), which is to make a preliminary prediction of whether the aneurysm will rupture. The network simply consists of convolutional layers and max-pooling layers.

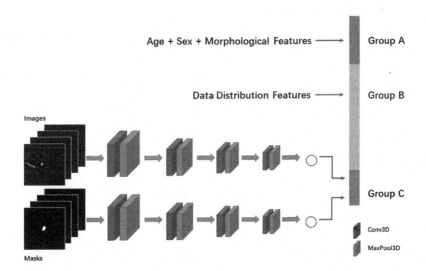

Fig. 2. The extracted features are divided into three groups. Group A contains age, sex and morphological features. Group B consists of data distribution features. The features extracted by CNN form the group C. Group A, B and C are concatenated as the input of the classification method.

2.2   Classification

XGBoost and the fully connected network are used to analyze and classify the extracted features to estimate the risk of cerebral aneurysm rupture.

3   Experiment

3.1   Details

After removing samples with missing information, we set 100 samples as the training set and 24 samples as the validation set. For each case, we numbered the extracted features and divided them into three groups. The group A includes age, sex, and morphological features which are numbered f[0–9]. The group B are data distribution features numbered f[10–107]. The features numbered f[108–109] extracted by the CNN whose structure is shown in Fig. 2 form the group C. Concerning the CNN extractor, we set the batch size to 4, total epochs to 100, and use SGD as the optimizer with a learning rate starting from 5e−3 and a momentum of 0.9.

In the XGBoost method, the number of estimator is set to 100, and the maximum depth is set to 5. The fully connected network has two hidden layers, each with 128 nodes. The SGD algorithm was used for optimization with an initial learning rate of 5e−3.

3.2   Results and Discussions

The *accuracy, recall, precision* and *F2-score* for different combinations of features and method are calculated, as Table 1 shows.

Table 1. The accuracy, recall, precision and F2-score for different combinations of features and method of validation set.

	Accuracy	Recall	Precision	F2 score
Group A + XGBoost	0.609	0.500	0.556	0.510
Group B + XGBoost	0.435	0.500	0.385	0.472
Group C + XGBoost	0.695	0.600	0.667	0.612
Group A, B + XGBoost	0.522	0.600	0.462	0.567
Group B, C + XGBoost	0.625	0.600	0.545	0.588
Group A, C + XGBoost	0.550	**0.800**	0.533	0.723
Group A, B, C + XGBoost	0.652	0.700	0.583	0.673
Group A + FCN	0.696	0.600	0.667	0.612
Group B + FCN	0.522	0.500	0.455	0.490
Group C + FCN	0.609	0.500	0.556	0.510
Group A, B + FCN	0.609	0.700	0.538	0.660
Group B, C + FCN	0.696	0.600	0.667	0.612
Group A, C + FCN	0.739	0.600	0.750	0.625
Group A, B, C + FCN	**0.826**	0.700	**0.875**	**0.729**

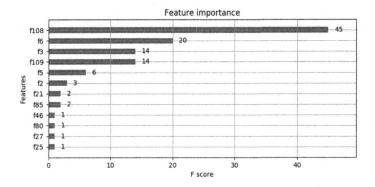

Fig. 3. The importance of different features.

By comparison, the cerebral aneurysm rupture risk estimation result of the fully connected network is slightly better than XGBoost. According to the analysis of XGBoost, the importance of features is shown in Fig. 3, which illustrates that group C and A are more important. The *f108* is the feature that extracted by CNN from the images and the *f6* is the ratio of maximum diameter to length. Meanwhile, as the number of features increases, the performance of XGBoost decreases, which might be caused by the disturbance from useless features.

4   Conclusion

In this work, we proposed a semi-automatic method for estimating the risk of rupture of cerebral. A total of 110 features extracted from aneurysm morphology, data distribution and a CNN-based method are used to fit a fully connected network and XGBoost. The experiments showed that the feature of cerebral aneurysm labels extracted by CNN and the ratio of maximum diameter to length have the greatest impact on estimating results. The method achieved an F2-score of 0.729 on the validation set and 0.862 on the test set of CADA 2020.

References

1. Beck, J., et al.: Difference in configuration of ruptured and unruptured intracranial aneurysms determined by biplanar digital subtraction angiography. Acta Neurochir. **145**(10), 861–865 (2003)
2. Greving, J.P., et al.: Development of the phases score for prediction of risk of rupture of intracranial aneurysms: a pooled analysis of six prospective cohort studies. Lancet Neurol. **13**(1), 59–66 (2014)
3. Korja, M., Kivisaari, R., Jahromi, B.R., Lehto, H.: Size and location of ruptured intracranial aneurysms: consecutive series of 1993 hospital-admitted patients. J. Neurosurg. **127**(4), 748–753 (2016)
4. Villablanca, J.P., et al.: Natural history of asymptomatic unruptured cerebral aneurysms evaluated at CT angiography: growth and rupture incidence and correlation with epidemiologic risk factors. Radiology **269**(1), 258–265 (2013)

5. Wiebers, D.O., et al.: Unruptured intracranial aneurysms: natural history, clinical outcome, and risks of surgical and endovascular treatment. Lancet **362**(9378), 103–110 (2003)
6. Yi, J., Zielinski, D., Chen, M.: Cerebral aneurysm size before and after rupture: case series and literature review. J. Stroke Cerebrovasc. Dis. **25**(5), 1244–1248 (2016)

Intracranial Aneurysm Rupture Risk Estimation Utilizing Vessel-Graphs and Machine Learning

Matthias Ivantsits[1]($\boxtimes$), Markus Huellebrand[1,2], Sebastian Kelle[1,3,4],
Titus Kuehne[1,3,4], and Anja Hennemuth[1,2,4]

[1] Charité – Universitätsmedizin Berlin, Augustenburger Pl. 1, 13353 Berlin, Germany
matthias.ivantsits@charite.de
[2] Fraunhofer MEVIS, Am Fallturm 1, 28359 Bremen, Germany
[3] German Heart Institute Berlin, Augustenburger Pl. 1, 13353 Berlin, Germany
[4] DZHK (German Centre for Cardiovascular Research), Berlin, Germany

Abstract. Intracranial aneurysms frequently cause subarachnoid hemorrhage—a life-threatening condition with a high mortality and morbidity rate. State-of-the-art methods combine demographic, clinical, morphological, and computational fluid dynamics parameters.

We propose a method combining morphological radiomics features, gray-level radiomics features, and a novel aneurysm site location encoding via directed graphs on the vessel tree. Some of the gray-level features seem to be good proxies for blood flow within the vessel and the aneurysms. Furthermore, our proposed method shows improved F2-scores and accuracy across various models fed with the aneurysm site encoding. A K-nearest neighbors method shows the best results during our model selection with an F2-score of 0.7 and an accuracy of 0.73 on the relatively small private test set with 22 individuals and 30 aneurysms.

Keywords: Intracranial aneurysms · Subarachnoid hemorrhage · X-ray rotational angiography · Machine learning · Rupture risks · Radiomics

1 Introduction

Cerebral aneurysms, also known as an intracranial aneurysms, are local dilations of blood vessels caused by a vessel walls' weakness. Subarachnoid hemorrhage (SAH) caused by an aneurysm rupture is a life-threatening condition with high mortality and morbidity. The death rate is above 40% [1], and in case of survival, cognitive impairment can affect patients for a long time, even lifelong. It is therefore highly desirable to detect aneurysms early and decide about the appropriate rupture prevention strategy.

Smoking is a crucial factor in predicting the rupture of cerebral aneurysms [2,3]. Patient-specific parameters like age, sex, and hypertension are additional factors in the rupture prediction [3]. Furthermore, there is a significant difference

© Springer Nature Switzerland AG 2021
A. Hennemuth et al. (Eds.): CADA 2020, LNCS 12643, pp. 93–103, 2021.
https://doi.org/10.1007/978-3-030-72862-5_10

in the patient's human geography predicting the outcome of a cerebral aneurysm [4]. The cerebral vessel network is formed by two vertebral and two internal carotid arteries. The location of the aneurysm within this thorny vessel tree is another decisive component in the assessment [5,6].

Widely-used high-resolution medical imaging such as magnetic resonance imaging (MRI), computed tomography (CT), or digital subtraction angiography (DSA) results in frequent detection of unruptured cerebral aneurysms [7]. From these image modalities, morphological features can be extracted to quantify the aneurysm's rupture risk. These shape-based features are significant factors in estimating the rupture of the aneurysm [3,4,8–11,14]. Furthermore, essential hemodynamic parameters can be inferred from the shape of the aneurysm and the surrounding vessel. Forces acting on aneurysm walls such as wall shear stress or oscillatory shear index have shown to be critical hemodynamic features [3–6,9–13].

Radiomics has proven to be a very promising toolkit for medical image processing, analysis, and interpretation. It derives vast amounts of features from imaged structures describing patterns in morphology and texture that are usually hard to differentiate with the bare eye. Initially, radiomics has been employed for oncological applications and an emerging technique in the cardiovascular field, especially with MRI. This observation has been confirmed by studies [15–19], which are extracting shape-based radiomics features and classifying diverse heart diseases.

We hypothesize that textural features derived from the aneurysm introduce valuable information to analyze the rupture potential. Furthermore, we propose a novel automatic method to encode the aneurysm position within the complex vessel network. These parameters are used along with patient demographic and shape-based parameters to assess the aneurysm rupture for the "Cranial aneurysm detection challenge" CADA - rupture risk estimation challenge [20].

2 Method and Materials

2.1 Dataset

The CADA dataset [20] was acquired utilizing the digital subtraction AXIOM Artis C-arm system with a rotational acquisition time of 5 s with 126 frames (190° or 1.5° per frame, 1024 × 1024-pixel matrix, 126 frames). Postprocessing was performed using LEONARDO InSpace 3D (Siemens, Forchheim, Germany). A contrast agent (Imeron 300, Bracco Imaging Deutschland GmbH, Germany) was manually injected into the internal carotid (anterior aneurysms) or vertebral (posterior aneurysms) artery. Reconstruction of a volume of interest selected by a neurosurgeon generated a stack of 440 image slices with matrices of 512 × 512 voxels in-plane, resulting in an iso-voxel size of 0.25 mm. The images were acquired in the Neurosurgery Department, Helios Klinikum Berlin-Buch. In total, the dataset consists of 131 rotational X-ray angiographic images from different patients. A private dataset of 22 images was held back as a test set by the CADA challenge organizers during the training period and only later released to allow offline processing and send in the results for centralized evaluation. The remaining 109 cases were available for model training and validation.

2.2   Feature Extraction

Our method consists of morphological and textural radiomics feature extraction [21]. The image can be used as direct input to a classification via a convolutional neural network or similar architectures, but since the image is a very high dimensional input, it is hard to train and prone to overfit if not parameterized correctly. Therefore, we argue that using a lower-dimensional input by deriving radiomics features produces a more reliable estimate of the aneurysm rupture. The radiomics feature classes we extract are: **First Order**, **Shape (3D)**, **Gray Level Co-occurrence Matrix (GLCM)**, **Gray Level Size Zone Matrix (GLSZM)**, **Gray Level Run Length Matrix (GLRLM)**, **Neighbouring Gray Tone Difference Matrix (NGTDM)**, and **Gray Level Dependence Matrix (GLDM)**. The shape-based features have shown to be useful for assessing cerebral aneurysm ruptures [15–19]. We hypothesize that the radiomics texture features approximate flow characteristics within the aneurysm and the surrounding vessel due to the injected contrast agent. The extracted radiomics features result in 108 variables for each aneurysm.

Furthermore, we calculate additional geometric features that have been significant in the rupture determination. These variables include the vessel curvature and diameter at the aneurysm, plus the maximum and minimum neck diameter of the aneurysm, which is illustrated in Fig. 1 on the right. We calculate the diameter on a window of 4 cm on this connected vessel. Furthermore, we calculate the curvature of this vessel on the same window. The window for the curvature at each point is set to 1 mm. Moreover, we consider supplementary features describing curvature and shape of the vessel dilation, including the **Gaussian curvature**, **convexity**, **curvedness**, **curvature**, and **PTypes**. The latter feature encodes each aneurysm mesh node as a peak, depression, or saddle point. An overview of these features is illustrated in Table 1.

For the positional encoding of the aneurysm within the vessel tree, we normalize the image by a percentile mapping. The 1- and 99-percentile of the image intensities are mapped to zero and one, respectively. The linear scaling is applied without clipping at the boundaries. Due to the small volume in relation to the background, the intensity of vessels and aneurysms is typically located above the 99-percentile, resulting in normalized values greater than 1. Next, we apply a threshold ≥ 1.5 followed by a morphological closing operation with a 4-neighborhood to segment the vessel tree. From this binarized image, we perform a skeletonization to extract the centerline of the vessel tree. Figure 1 illustrates the extracted centerline of this described process, which is internally represented as an undirected graph. Next, we find the point in this graph that is furthest down in the z-dimension, the carotid or vertebral artery. We define this as the new root of the graph and convert it to a directed non-cyclic graph. Based on this representation we extract the path from the root of the carotid/vertebral artery to the aneurysm and calculate the **number of bifurcations**, **vessel length**,

Table 1. An overview of the additional features used during the sequential feature selection process.

Feature	Description
Vessel curvature (vessel)	The curvature of the vessel connected to the aneurysm. Including the minimum, maximum, std-dev, and mean of the curvature
Vessel diameter (vessel)	The diameter of the vessel connected to the aneurysm. Including the minimum, maximum, std-dev, and mean of the diameter
Neck diameter (shape)	The neck diameter of the aneurysm. Including the minimum, maximum
CGauss (shape)	The Gaussian curvature of the aneurysm. Including the minimum, maximum, std-dev, and mean
Convexity (shape)	The convexity of the aneurysm. Including the minimum, maximum, std-dev, and mean
Curvedness (shape)	The Gaussian curvedness of the aneurysm. Including the minimum, maximum, std-dev, and mean
MinCurvature (shape)	The minimum principal curvature of the aneurysm (k_2). Including the minimum, maximum, std-dev, and mean
MaxCurvature (shape)	The maximum principal curvature of the aneurysm (k_1p). Including the minimum, maximum, std-dev, and mean
MeanCurvature (shape)	The mean curvature of the aneurysm. Including the minimum, maximum, std-dev, and mean
PTypes (shape)	The node type of each node point of the aneurysm, where one describes a peak point, two a depression point, and three a saddle point
#bifurcations (path)	The number of bifurcations from the carotis to the aneurysm
Bifurcation-0 (path)	One if the aneurysm is located in the larger vessel sub-tree at the first bifurcation, zero otherwise
Bifurcation-1 (path)	One if the aneurysm is located in the larger vessel sub-tree at the second bifurcation, zero otherwise
Bifurcation-2 (path)	One if the aneurysm is located in the larger vessel sub-tree at the third bifurcation, zero otherwise
Vessel length (path)	The path length from the carotis to the aneurysm
Vessel volume (path)	The vessel volume of the path from the carotis to the aneurysm

vessel volume, and **bifurcation-x**. Where the **number of bifurcations** is the number of bifurcations from the root to the aneurysm. The **vessel length** and **vessel volume** are the length and volume from the root to the aneurysm, respectively. **Bifurcation-x** represents three binary variables for the first three bifurcations. Where it is set to zero if the bifurcation leading to the aneurysm is in the smaller volume of the two sub-trees and one otherwise.

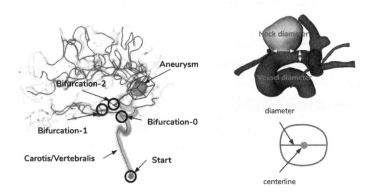

Fig. 1. An illustration of the path encoding features calculated on the centerline of the vessel tree on the **left**. **Right** exemplifies the **neck diameter** and the **vessel diameter**.

2.3   Classification

For the classification of ruptured versus non-ruptured aneurysms, we decided to cross-validate five distinct models. Support vector machines (SVM) are particularly useful in high dimensional spaces with a clear margin between the classes and have shown to work well on pathological case detection by Cetin et al. [22]. Random forests are very robust to outliers and comparatively little impacted by noise. They have shown to be effective in medical settings, as illustrated by [23–25]. Multilayer perceptrons (MLP) have been thoroughly studied, and Isensee et al. [25] have illustrated that they can be successfully applied for pathological case detection. AdaBoost is one of many ensemble methods, which utilizes a collection of weakly trained classifiers. It is useful in individual treatment estimations [26]. The K-nearest neighbors (KNN) algorithm does not need any training, and therefore new data can be seamlessly added. Akhil [27] has demonstrated medical applicability by predicting various heart diseases (Table 2).

Table 2. Five models and their respective hyperparameters used during the 8-fold nested cross-validation.

Model name	Hyperparameters
Support vector machine	C: 0.01, 0.05, 0.1, 1 kernel: linear, poly, rbf, sigmoid gamma: auto, scale
Random forest	n_estimators: 100, 200, 300 max_depth: 90, 100, 110 max_features: 2, 3 min_samples_leaf: 3, 4, 5
AdaBoost	n_estimators: 10, 20, 50
Multilayer perceptron	hidden_layer_sizes: (100,), (100, 50), (50,), (50, 25) max_iter: 200, 300, 500, 700
K-nearest neighbors	n_neighbors: 5–20 weights: uniform, distance metric: minkowski, euclidean, manhattan

3   Results

We performed all experiments on an Intel(R) Core(TM) i7-8700K CPU @ 3.70 GHz with 16 GBs RAM. Since the contest is evaluated on the predictions' F2-score, we evaluate all models based on this metric. We argue that a false-negative classification is more harmful to the screening process, and since the dataset is imbalanced, we proceed with this metric. The F2-score combines sensitivity and precision and considers sensitivity twice as important as precision. Before the model fits, the features are normalized by the min-max of the training set.

The first experiment we conducted, was a sequential feature selection (SFS) on all described features mentioned in Sect. 2.2, excluding the path encoding variables. SFS is a greedy search algorithm to reduce the input dimensionality. Furthermore, we impose a floating operation on the SFS algorithm, which allows already selected features to be replaced at a later stage of the method. We perform a 5-fold cross-validation (CV) on all models and features to select the final set of variables. Due to studies implying the importance of morphological features [3,4,8–11,14] to estimate the risk of rupture, we fix several features during the SFS procedure. These features are definitely selected in the final subset and include the following variables: **sphericity, mesh-volume, minor-axis-length, major-axis-length, least-axis-length, elongation, flatness, surface area, patient sex**, and **patient age**.

The results of this analysis are illustrated in Fig. 2. It presents the mean F2-score versus the number of added features (k), including the 95% confidence interval of the best-selected model during the CV. Since the CV on all models and features is an exhaustive search which takes roughly 36 h on the described hardware, we opted to implement a voting mechanism of all selected features

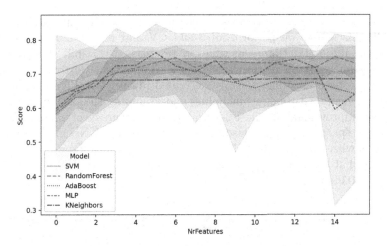

Fig. 2. The mean F2-score for the best CV model versus the number of features added. This includes the 95% confidence interval.

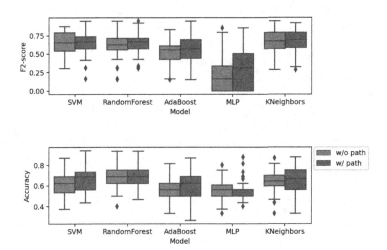

Fig. 3. A comparison of the baseline model including 20 features from our voting method and this set of features extended by the path encoding. The models trained with the path encoding show superior results in the F2-score as well as the accuracy.

across the models. For this voting schema, we score each feature with the inverse rank divided by k, where k is the number of features with the maximum F2-score (e.g., for the SVM $k = 14$). Next, we average overall scores and select the top 20 features for our final model.

We selected the model with these 20 features as a baseline in the next experiment. In this setup, we perform an 8-fold CV and repeat it ten times with different training/validation seeds to avoid any bias on the split and get a more accurate estimate of the performance. As a comparison, we train all models, including the path encoding features described in Sect. 2.2. Figure 3 illustrates these experiments with models including and excluding the path encoding. We exhibit an improvement in the F2-score and the accuracy of all models when including the path encodings. Based on the F2-score, the KNN model shows the best results with a mean of 0.69 ± 0.13. Evaluating the accuracy's performance, we observe that the random forest is slightly superior to the KNN with a mean accuracy of 0.69 ± 0.09 compared to 0.66 ± 0.11.

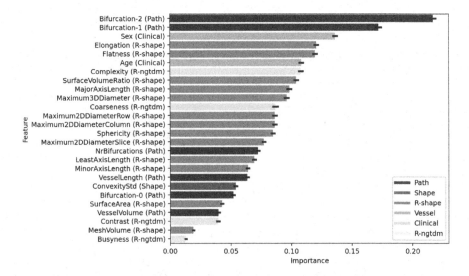

Fig. 4. The feature importance scores on the finally trained KNN, where the importance is defined by the difference of the models' baseline and the score after a feature permutation.

Since the challenge is evaluated on the F2-score, we chose the KNN model to be the best method. The model is initialized with **15 neighbors, distance weights**, and the **Minkowski metric**. The **distance weights** are weighted with the inverse distance to other points. For the **Minkowski metric**, we chose a power parameter of two, which corresponds to the standard Euclidean metric. Consequently, we performed a feature importance analysis as proposed by Breiman [28]. This analysis can be performed on any fitted model by calculating a base score produced by the training or test set model. This is followed by a random shuffle to one of the features and compared to the baseline's predictive power. This procedure is then repeated and applied to all features to come up with an importance score. The feature importance scores of this model are illustrated in Fig. 4. We observe high importance for some of the path encoding features, which corresponds to the importance of the aneurysm site mentioned in the literature and highlights this method's potential. Furthermore, this analysis shows very high importance for the clinical variables (**age** and **sex**). Some of the gray-level features calculated from the **NGTDM** matrix are ranked high in this assessment. This includes the **complexity**, which describes the uniformity of the intensities within the aneurysm. For the clinical integration of the proposed method, we utilized LIME [29], which is an explanatory framework for any black-box classifier.

4   Discussion and Conclusion

We have illustrated an aneurysm rupture risk estimation pipeline, which includes calculating vast amounts of radiomics features. Furthermore, we extend these features by multiple other morphological features, features describing the vessel, and a location encoding of the aneurysm. Our proposed method includes a sequential feature selection to reduce the dimensionality of the input. This is followed up by a nested cross-validation to fit different classifiers. The KNN model with **15 neighbors, distance weights**, and the **Minkowski metric** shows the best results on the F2-score of 0.69 ± 0.13 on the CV and 0.7 F2-score and 0.73 accuracy on the private test set. The discrepancy of the F2-score on the CV and private test set lies outside of the 95% standard error interval. This is partly due to the small test set and the difference of the rupture to non-rupture ratio within the training (0.44) and the private test set (0.37).

Interestingly the KNN model outperforms the other models, in most cases. This is probably due to the few hyperparameters to choose from, limiting the overfitting of the model. Furthermore, KNNs are particularly better with few features than the SVM, which accelerates the performance with a high dimensional input. The model includes 25 morphological, intensity-based, and features describing the site of the vessel dilation. The inclusion of the path encoding improves the models' predictive power by 0.04 on the discussed metric. These features are ranked according to the final models' importance for clinical interpretation (Fig. 4). The importance of the **NGTDM** features implies that they can approximate flow patterns within the vessel dilation. Nonetheless, this needs to be verified with a larger dataset due to the vast dimensionality and low sample size.

Mentionable is that neither the vessel diameter/curvature nor the neck diameter was considered essential during the sequential feature selection. These features probably need to be extended by the angle between the vessel and the aneurysm [30]. The combination of these features gives improved approximations of the resulting jet and thus stress on the vessel. Furthermore, the method can be improved by a more elaborate site location identification. This can either be done by a convolutional neural network applied to the image or by a point cloud classification into anatomical sections by a graph neural network based on the vessel tree's centerline. Lastly, hemodynamical parameters are of the essence to estimate the risk of rupture.

References

1. Teunissen, L.L., et al.: Risk Factors for Subarachnoid Hemorrhage (1996)
2. Can, A., et al.: Association of intracranial aneurysm rupture with smoking duration, intensity, and cessation (2017)
3. Chabert, S., et al.: Applying machine learning and image feature extraction techniques to the problem of cerebral aneurysm rupture (2017)
4. Detmer, F.J., et al.: Extending statistical learning for aneurysm rupture assessment to Finnish and Japanese populations using morphology, hemodynamics, and patient characteristics (2019)
5. Cebral, J.R., et al.: Analysis of hemodynamics and wall mechanics at sites of cerebral aneurysm rupture (2015)
6. Detmer, F.J., et al.: Associations of hemodynamics, morphology, and patient characteristics with aneurysm rupture stratified by aneurysm location (2019)
7. Thompson, B.G., et al.: Guidelines for the management of patients with unruptured intracranial aneurysms a guideline for healthcare professionals from the American heart association/American stroke association (2015)
8. Lindgren, A.E., et al.: Irregular shape of intracranial aneurysm indicates rupture risk irrespective of size in a population-based cohort (2016)
9. Tanioka, S., et al.: Machine learning classification of cerebral aneurysm rupture status with morphologic variables and hemodynamic parameters (2020)
10. Paliwal, N., et al.: Outcome prediction of intracranial aneurysm treatment by flow diverters using machine learning (2018)
11. Xiang, J., et al.: Hemodynamic–morphologic discriminants for intracranial aneurysm rupture (2011)
12. Suzuki, M., et al.: Classification model for cerebral aneurysm rupture prediction using medical and blood-flow-simulation data (2019)
13. Chen, G., et al.: Development and validation of machine learning prediction model based on computed tomography angiography-derived hemodynamics for rupture status of intracranial aneurysms: a Chinese multicenter study (2020)
14. Kleinloog, R., de Mul, N., Verweij, B.H., Post, J.A., Rinkel, G.J.E., Ruigrok, Y.M.: Risk Factors for intracranial aneurysm rupture: a systematic review (2018)
15. Chandra, A.R., et al.: Initial study of the radiomics of intracranial aneurysms using Angiographic Parametric Imaging (API) to evaluate contrast flow changes (2019)
16. Podgorsak, A.R., et al.: Automatic radiomic feature extraction using deep learning for angiographic parametric imaging of intracranial aneurysms (2020)
17. Liu, Q., Jiang, P., Jiang, Y., Li, S., Ge, H., Jin, H., Li, Y.: Bifurcation configuration is an independent risk factor for aneurysm rupture irrespective of location (2019)
18. Liu, Q., et al.: Prediction of aneurysm stability using a machine learning model based on pyradiomics-derived morphological features (2019)
19. Juchler, N., et al.: Radiomics approach to quantify shape irregularity from crowd-based qualitative assessment of intracranial aneurysms (2020)
20. CADA rupture risk estimation challenge. https://cada-rre.grand-challenge.org/. Accessed 05 Oct 2020
21. van Griethuysen, J.J.M., et al.: Computational radiomics system to decode the radiographic phenotype (2017)
22. Cetin, I.: A radiomics approach to computer-aided diagnosis with cardiac Cine-MRI (2017)
23. Wolterink, J.M., Leiner, T., Viergever, M.A., Išgum, I.: Automatic segmentation and disease classification using cardiac cine MR images (2017)

24. Khened, M., Alex, V., Krishnamurthi, G.: Densely connected fully convolutional network for short-axis cardiac cine MR image segmentation and heart diagnosis using random forest (2017)
25. Isensee, F., Jaeger, P.F., Full, P.M., Wolf, I., Engelhardt, S., Maier-Hein, K.H.: Automatic cardiac disease assessment on cine-MRI via time-series segmentation and domain specific features (2017)
26. Sugasawa, S., Noma, H.: Estimating individual treatment effects by gradient boosting trees (2019)
27. Akhil, J.: Prediction of heart disease using k-nearest neighbor and particle swarm optimization (2017)
28. Leo Breiman, Random Forests (2001)
29. Túlio Ribeiro, M., Singh, S., Guestrin, C.: Why should i trust you? Explaining the predictions of any classifier (2016)
30. Jiang, P.: A novel scoring system for rupture risk stratification of intracranial aneurysms: a hemodynamic and morphological study (2018)

Intracranial Aneurysm Rupture Prediction with Computational Fluid Dynamics Point Clouds

Matthias Ivantsits[1,3](✉), Leonid Goubergrits[1,3,6], Jan Brüning[1,3],
Andreas Spuler[3,5], and Anja Hennemuth[1,2,3,4]

[1] Charité – Universitätsmedizin Berlin, Augustenburger Pl. 1, 13353 Berlin, Germany
matthias.ivantsits@charite.de
[2] Fraunhofer MEVIS, Am Fallturm 1, 28359 Bremen, Germany
[3] German Heart Institute Berlin, Augustenburger Pl. 1, 13353 Berlin, Germany
[4] DZHK (German Centre for Cardiovascular Research), Berlin, Germany
[5] Helios Hospital Berlin-Buch, Schwanebecker Chaussee 50, 13125 Berlin, Germany
[6] Einstein Center Digital Future, Wilhelmstraße 67, 10117 Berlin, Germany

Abstract. Intracranial aneurysms frequently cause subarachnoid hemorrhage—a life-threatening condition with a high mortality and morbidity rate. State-of-the-art methods of the rupture risk prediction combine demographic, clinical, morphological, and computational fluid dynamics based hemodynamic parameters. We propose a method of blending morphological features, computational fluid dynamics parameters, and patient demographic features. The shape and wall-shear-stress at each point of the aneurysm are encoded with a deep point cloud neural network and extended by additional location encodings of the aneurysm as well as age and sex of the patient. On this concatenated feature vector, an MLP infers the probability of rupture for a given cerebral aneurysm. The proposed network was trained on the **CADA - rupture risk estimation** challenge set of 109 aneurysms. The proposed method achieves an accuracy of 0.64 and an F2-score of 0.73 on the private **CADA - rupture risk estimation** challenge test set of 30 aneurysms.

Keywords: Intracranial aneurysms · Subarachnoid hemorrhage · X-ray rotational angiography · Machine learning · Rupture risks · Deep learning

1 Introduction

Intracranial aneurysms, are local dilatations of blood vessels caused by vessel walls' weakness. Subarachnoid hemorrhage (SAH) caused by an aneurysm rupture is a life-threatening condition with high mortality and morbidity. Although, the rupture rate is very low, the death rate at rupture is above 40% [1], and in case of survival, cognitive impairment can affect patients for a long time, even lifelong. It is therefore highly desirable to detect aneurysms early and decide about the appropriate rupture prevention strategy.

© Springer Nature Switzerland AG 2021
A. Hennemuth et al. (Eds.): CADA 2020, LNCS 12643, pp. 104–112, 2021.
https://doi.org/10.1007/978-3-030-72862-5_11

Smoking is a crucial factor in predicting the rupture of cerebral aneurysms [2,3]. Patient-specific parameters like age, sex, and hypertension are additional factors in the rupture prediction [3]. Furthermore, there is a significant difference in the patient's geography predicting the outcome of a cerebral aneurysm [4]. The cerebral vessel network is formed by two vertebral and two internal carotid arteries. The location of the aneurysm within this thorny vessel tree is another decisive component in the assessment [5,6].

Widely-used high-resolution medical imaging such as magnetic resonance imaging (MRI), computed tomography (CT), or digital subtraction angiography (DSA) results in frequent detection of unruptured cerebral aneurysms [7]. From these image modalities, morphological features can be extracted to quantify the aneurysm's rupture risk. These shape-based features are significant factors in estimating the rupture of the aneurysm [3,4,8–11,14]. Furthermore, essential hemodynamic parameters can be inferred from the shape of the aneurysm and the surrounding vessel. Forces acting on aneurysm walls such as wall shear stress or oscillatory shear index have shown to be critical hemodynamic features [3–6,9–13].

Radiomics has proven to be a very promising toolkit for medical image processing, analysis, and interpretation. It derives vast amounts of features from imaged structures describing patterns in morphology and texture that are usually hard to differentiate with the bare eye. Studies have confirmed this observation [15–19], which are extracting shape-based radiomics features and classifying diverse heart diseases. Nonetheless, this feature extraction results in a compressed representation of the data distribution and lacks descriptions of the detailed morphology. Recent developments in deep learning utilizing point clouds are promising methods to classify objects based on their shape and additional features associated with the 3D coordinates. Initially, this point cloud data was rendered in regularized 3D grids, which results in unnecessarily voluminous data representation. This inflated representation demands more extensive networks with many more parameters to fully capture the underlying data distribution. The application of point-cloud classifications in medical settings [20,21] has shown to be successful and is a promising approach for disease classification based on morphological patterns.

2   Method and Materials

2.1   Dataset

The CADA dataset [24] was acquired utilizing the digital subtraction AXIOM Artis C-arm system with a rotational acquisition time of 5 s with 126 frames (190° or 1.5° per frame, 1024 × 1024-pixel matrix, 126 frames). Postprocessing was performed using LEONARDO InSpace 3D (Siemens, Forchheim, Germany). A contrast agent (Imeron 300, Bracco Imaging Deutschland GmbH, Germany) was manually injected into the internal carotid (anterior aneurysms) or vertebral (posterior aneurysms) artery. Reconstruction of a volume of interest selected by a neurosurgeon generated a stack of 440 image slices with matrices of

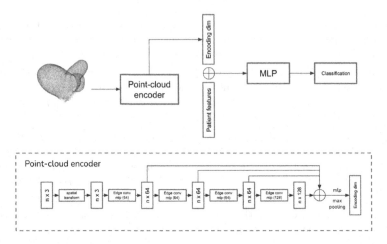

Fig. 1. An illustration of the proposed architecture. On top, the overall architecture with the point-cloud encoding and the concatenation of additional features is shown. Based on this concatenated feature vector, an MLP is trained to predict the final rupture status. On the bottom, the point-cloud encoding network is illustrated. This encoding is comprised of multiple k-point samplings and convolutions.

512×512 voxels in-plane, resulting in an iso-voxel size of 0.25 mm. The images were acquired in the Neurosurgery Department, Helios Klinikum Berlin-Buch. In total, the dataset consists of 131 rotational X-ray angiographic images from different patients. A private dataset of 22 images was held back as a test set by the CADA challenge organizers during the training period and only later released to allow offline processing and send in the results for centralized evaluation. The remaining 109 cases were available for model training and validation.

2.2   Method

CFD Analysis of Hemodynamics. Commercial software Fluent (ANSYS-Fluent Inc. Canonsburg, Pennsylvania, USA) was used to solve the Navier-Stokes equations for mass and momentum conservation simulating controlled cycle-averaged flow conditions. Laminar flow was assumed. A triangulated surface mesh was generated using a radius-dependent resolution of 0.15–0.25 mm of node distance as a resolution corresponding to the 3DRA imaging modality resolution. Next, a volume mesh including a boundary mesh layer consisting of three rows of wedges with an increasing depth of wedges by a factor of 1.2 (total depth of 0.3 mm) refining the mesh in the near-wall region was generated to warrant the accurate assessment of the wall shear stress according to the mesh independence study. Near wall flow assessment with CFD was earlier validated against in vitro experiments [28]. Blood was modeled as non-Newtonian fluid using the generalized power-law model with a patient-specific hematocrit value [29]. Since clinicians provide only imaging data of anatomy, patient-specific boundary conditions for inlets and outlets were not available. The averaged flow rate of the internal carotid artery of 222 ml/min was set according to literature data [30] for all aneurysms of the anterior circulation, whereas all aneurysms of the posterior circulation were simulated with the averaged flow rate of the vertebral artery of 66.5 ml/min [31]. Plug inlet velocity profile was set. Flow rate splitting for all outlets was calculated by using area-dependent weighting [32]. The vessel and aneurysm walls were assumed to be rigid, and a no-slip condition at the wall was set. A large number of hemodynamic parameters was proposed as aneurysm rupture risk predictors. The majority of these parameters are summarized here [33]. However, as shown by the analysis of the rupture risk assessment in frames of the Multiple Aneurysm AnaTomy CHallenge (MATCH) 2018, the most often used hemodynamic rupture risk parameter was the wall shear stress [34].

Point-Cloud Classification. For the prediction of ruptured vs. non-ruptured aneurysms, we propose a point-cloud encoding network. This encoding is concatenated with additional patient-specific and location features. An overview of the proposed method is illustrated in Fig. 1. To encode the aneurysm morphology, we suggest a dynamic graph convolutional neural network (CNN) introduced by Wang et al. [22]. The proposed method is inspired by the PointNet architecture by Qi et al. [23]. Wang et al. exploit the local geometry of a point by sampling k-nearest neighbors of each point followed by a convolutional operation. Furthermore, we extend each vector's three spatial dimensions by an additional dimension describing the wall shear stress (WSS) at each sampled point.

Moreover, we extend the defined morphological and WSS encoding by supplementary patient features. These features are comprised of the patients **age** and **sex**. Furthermore, we experiment with a location encoding of the aneurysm within the complex vessel-tree. This encoding is defined by the arteries within this vessel-tree and include - the **anterior communicating artery (Acom)**, **posterior communicating artery (Pcom)**, **posterior inferior cerebellar artery (PICA)**, **anterior cerebral artery (ACA)**, **middle cerebral artery (MCA)**, **internal carotid artery (ICA)**, **basilar artery (Bas)**, **anterior choroidal artery (AChA)**, **internal carotid artery terminus (ICAT)**, and **posterior cerebral artery (PCA)**. These locations are one-hot encoded for the representation of the neural network.

The aneurysm shape plus WSS encoding is concatenated with the location encoding and the patient features in a final step. This feature vector is the input to a final fully connected neural network to estimate the risk of rupture. Furthermore, we utilize some data augmentation techniques. These augmentations include small random jitter to the WSS features and random rotations around the X-, Y-, and Z-axis of $\pm 15°$.

3 Results

We performed all experiments on an Intel(R) Core(TM) i7-8700K CPU @ 3.70 GHz with 16 GBs RAM and an Nvidia RTX 2080 Ti GPU with 10 GB memory. Since the contest is evaluated on the predictions' F2-score, we evaluate all models based on this metric. We argue that a false-negative classification is more harmful to the screening process, and since the dataset is imbalanced, we proceed with this metric. The F2-score combines sensitivity and precision and considers sensitivity twice as important as precision. Before the model fits, the features are normalized by the min-max of the training set.

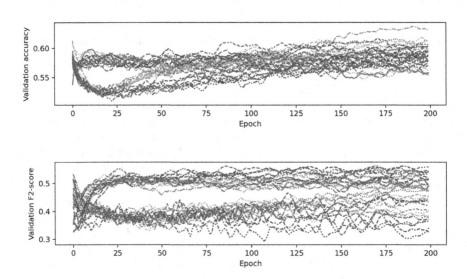

Fig. 2. An illustration of the performed grid-search performed with a 5-fold validation.

The first experiment we conducted was a 5-fold cross validation. We repeated this validation procedure five times with different splits to remove any bias during the splits. For this experiment we performed a grid-search to find good hyperparameters for the following tests. The results of this examination is illustrated in Fig. 2. This grid-search was conducted for the **batch-size**, the **optimizer**, **scheduler**, the **encoding-dimension**, and the **number of points** sampled from the input mesh. For the **batch-size** we tested for 16 and 32, for the **optimizer** Adam and SGD, for the **scheduler** a cosine and step scheduling schema, for the **encoding-dimensions** 512 and 1024, and for the **number of points** sampled from the input mesh 512 and 1024. A **batch-size** of 32, the Adam **optimizer**, a step **scheduling**, and 1024 for the **encoding-dimension** and the **number of points** sampled from the input mesh are optimal.

Figure 3 illustrates the results of the following experiments we conducted. Again we decided to evaluate the setup on a 5-fold cross-validation, which was

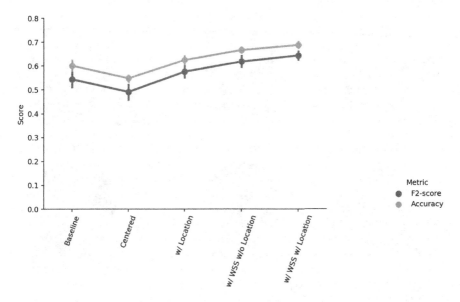

Fig. 3. An illustration of five different experimental setups. The baseline infers the rupture status only based on the morphology of the aneurysm plus the patients age and sex. The centered setup centers the aneurysm around its origin. The **w/ Location** setup adds the locations encoding. For the last two setups the WSS magnitude is added.

repeated five times. It highlights the accuracy and F2-score for five different experimental setups. The **baseline** was performed with the point-cloud, the patients **age**, and **sex** as input to the architecture. For the **centered** experiment, we used the same input and centered the aneurysm mesh around the origin. This setup decreased the model's performance by a few percentage points. Next, we included the location encoding as described in Sect. 2.2 as additional input. This analysis shows an improvement of the model's performance by 0.03 in accuracy compared to the baseline. For the fourth set up, we added the wall-shear-stress to each vector of the point-cloud. This setup increases the model's performance. For the final experiment, we utilized the same setup as in the previous exploration but added the location encoding, which again improves the performance by a few percentage points.

Figure 4 exemplary shows WSS distribution of the three ICA aneurysms of the test cohort. Three cases of the same ICA location represent true-positive, true-negative and false-positive rupture risk predictions. Wrong negative is not shown since among the test cohort no one anterior circulation case with wrong negative prediction was found.

Lastly, we trained ten models with different random seeds and created an ensemble to remove any negative or positive outliers. For the ensemble, we opted for a simple averaging of the probabilities and a threshold of 0.5. On the private **CADA - rupture risk estimation** challenge test set [24] this method achieves an accuracy of 0.64 and an F2-score of 0.73.

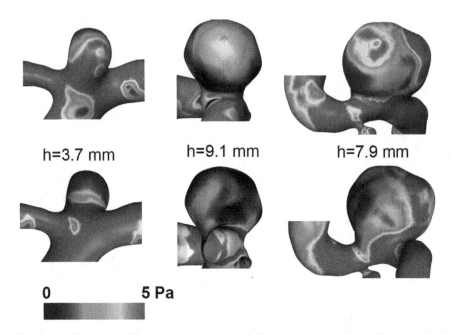

h=3.7 mm   h=9.1 mm   h=7.9 mm

0   5 Pa

Fig. 4. An illustration of WSS distribution of the three ICA aneurysms. From left to write: right positive, right negative and wrong positive rupture predictions. Upper row shows face wall, whereas bottom row shows back wall. Aneurysm size is represented by an aneurysm height h.

4   Discussion and Conclusion

We have illustrated a feasibility of the proposed cerebral aneurysm rupture prediction method within the top solutions of the **CADA - rupture risk estimation** challenge [24]. Remarkable is the drop in performance from the **baseline** to the **centered** experiment we conducted. It seems to be that the non-centered aneurysm meshes contain valuable implicit information on which vessel the dilation is located at. Furthermore, the increase in performance when including the location encoding is noticeable. This improvement confirms recent studies on the importance of the site the aneurysms are located at. Most notable is the performance boost after adding the wall-shear-stress to each input point. Again this is an indication that this CFD parameter introduces valuable information to the model. Considering our results in light of the previous aneurysm challenge MATCH 2018 [34] follow improvements in future are thinkable: incorporation of more hemodynamic parameters such as oscillating shear index or stagnation points, additional incorporation of morphometric parameters or features such as aspect ratio or daughter blebs. Furthermore, known but not available in the current challenge patient-related risk factors such as smoking, diabetes, hypertension, and earlier SAH events could also improve the performance of the rupture prediction.

To improve the proposed deep learning method, it is very likely that an increase in sample size can help better to predict the probability of rupture of a cerebral aneurysm. Current training set include only 8 aneurysms of the posterior circulation, which are known to be associated with higher rupture risk [35]. Among three posterior circulation aneurysms (all three ruptured) of the test dataset, one was predicted correctly and two wrong. An extension of the dataset as well as a good balance of data are crucial to predict the aneurysm rupture risk. The Aneurisk [25], Aneux [26], and DCAG [27] datasets would be valuable enrichments for further studies. Furthermore, a combination of radiomics features and point-cloud classifications might bring interesting insights to estimating aneurysm rupture risks.

References

1. Teunissen, L.L., et al.: Risk Factors for Subarachnoid Hemorrhage (1996)
2. Can, A., et al.: Association of intracranial aneurysm rupture with smoking duration, intensity, and cessation (2017)
3. Chabert, S., et al.: Applying machine learning and image feature extraction techniques to the problem of cerebral aneurysm rupture (2017)
4. Detmer, F.J., et al.: Extending statistical learning for aneurysm rupture assessment to Finnish and Japanese populations using morphology, hemodynamics, and patient characteristics (2019)
5. Cebral, J.R.: Analysis of hemodynamics and wall mechanics at sites of cerebral aneurysm rupture (2015)
6. Detmer, F.J., et al.: Associations of hemodynamics, morphology, and patient characteristics with aneurysm rupture stratified by aneurysm location (2019)
7. Thompson, B.G., et al.: Guidelines for the Management of Patients With Unruptured Intracranial Aneurysms A Guideline for Healthcare Professionals From the American Heart Association/American Stroke Association (2015)
8. Lindgren, A.E., et al.: Irregular shape of intracranial aneurysm indicates rupture risk irrespective of size in a population-based cohort (2016)
9. Tanioka, S., et al.: Machine learning classification of cerebral aneurysm rupture status with morphologic variables and hemodynamic parameters (2020)
10. Paliwal, N., et al.: Outcome prediction of intracranial aneurysm treatment by flow diverters using machine learning (2018)
11. Xiang, J., et al.: Hemodynamic–morphologic discriminants for intracranial aneurysm rupture (2011)
12. Suzuki, M., et al.: Classification model for cerebral aneurysm rupture prediction using medical and blood-flow-simulation data (2019)
13. Chen, G., et al.: Development and validation of machine learning prediction model based on computed tomography angiography-derived hemodynamics for rupture status of intracranial aneurysms: a Chinese multicenter study (2020)
14. Kleinloog, R., et al.: Risk factors for intracranial aneurysm rupture: a systematic review (2018)
15. Chandra, A.R., et al.: Initial study of the radiomics of intracranial aneurysms using Angiographic Parametric Imaging (API) to evaluate contrast flow changes (2019)
16. Podgorsak, A.R., et al.: Automatic radiomic feature extraction using deep learning for angiographic parametric imaging of intracranial aneurysms (2020)

17. Liu, Q., et al.: Bifurcation configuration is an independent risk factor for aneurysm rupture irrespective of location (2019)
18. Liu, Q., et al.: Prediction of aneurysm stability using a machine learning model based on pyradiomics-derived morphological features (2019)
19. Juchler, N., et al.: Radiomics approach to quantify shape irregularity from crowd-based qualitative assessment of intracranial aneurysms (2020)
20. Yang, L., Chakraborty, R.: A GMM based algorithm to generate point-cloud and its application to neuroimaging (2019)
21. Gutierrez-Becker, B., Wachinger, C.: Deep multi-structural shape analysis: application to neuroanatomy (2018)
22. Wang, Y., Sun, Y., Liu, Z., Sarma, S., Bronstein, M., Solomon, J.: Dynamic graph CNN for learning on point clouds (2018)
23. Ruizhongtai Qi, C., Su, H., Mo, K., Guibas, L.: PointNet: deep learning on point sets for 3D classification and segmentation (2016)
24. CADA rupture risk estimation challenge. https://cada-rre.grand-challenge.org/. Accessed 05 Oct 2020
25. AneuRisk dataset. http://ecm2.mathcs.emory.edu/aneuriskweb/repository. Accessed 26 Nov 2020
26. Aneux dataset. https://www.aneux.ch/home/internal/. Accessed 26 Nov 2020
27. Database of Cerebral Artery Geometries including Aneurysms at the Middle Cerebral Artery Bifurcation. https://figshare.shef.ac.uk/articles/dataset/Database_of_Cerebral_Artery_Geometries_including_Aneurysms_at_the_Middle_Cerebral_Artery_Bifurcation/4806910/1. Accessed 26 Nov 2020
28. Goubergrits, L., et al.: In vitro study of near-wall flow in a cerebral aneurysm model with and without coils (2010)
29. Wellnhofer, E., Osman, J., Kertzscher, U., Affeld, K., Fleck, E., Goubergrits, L.: Flow simulation studies in coronary arteries—impact of side-branches (2010)
30. Scheel, P., Ruge, Ch., Petruch, U.R., Schoening, M.: Color duplex measurement of cerebral blood flow volume in healthy adults (2000)
31. Kato, T., Indo, T., Yoshida, E., Iwasaki, Y., Sone, M., Sobue, G.: Contrast-enhanced 2D cine Phase MR angiography for measurement of basilar artery blood flow in posterior circulation ischemia (2002)
32. Cebral, J.R., Castro, M.A., Putman, C.M., Alperin, N.: Flow–area relationship in internal carotid and vertebral arteries (2008)
33. Goubergrits, L., Schaller, J., Kertzscher, U., Woelken, T., Ringelstein, M., Spuler, A.: Hemodynamic impact of cerebral aneurysm endovascular treatment devices: coils and flow diverters (2014)
34. Berg, P., et al.: Multiple Aneurysms AnaTomy CHallenge 2018 (MATCH)—phase II: rupture risk assessment (2019)
35. Wermer, M.J.H., van der Schaaf, I.C., Algra, A., Rinkel, G.J.E.: Risk of rupture of unruptured intracranial aneurysms in relation to patient and aneurysm characteristics (2007)

Correction to: Cerebral Aneurysm Detection and Analysis

Anja Hennemuth⦿, Leonid Goubergrits ⦿, Matthias Ivantsits,
and Jan-Martin Kuhnigk

Correction to:
A. Hennemuth et al. (Eds.): *Cerebral Aneurysm Detection,*
LNCS 12643, https://doi.org/10.1007/978-3-030-72862-5

In the originally published version of the book the title was incorrect. The title has now been corrected from "Cerebral Aneurysm Detection" to "Cerebral Aneurysm Detection and Analysis" and the publication has been updated.

The updated version of the book can be found at
https://doi.org/10.1007/978-3-030-72862-5

Author Index

Printed in the United States
by Baker & Taylor Publisher Services